# NONPOLLUTING COATINGS
# AND COATING PROCESSES

# NONPOLLUTING COATINGS AND COATING PROCESSES

Proceedings of an ACS Symposium held
August 30-31, 1972, in New York City

Edited by

## J. L. Gardon

*M & T Chemicals Inc.
Rahway, New Jersey*

*and*

## Joseph W. Prane

*Consultant
Elkins Park, Pennsylvania*

PLENUM PRESS • NEW YORK – LONDON • 1973

Library of Congress Catalog Card Number 72-97719

ISBN-13: 978-1-4684-0738-9     e-ISBN-13: 978-1-4684-0736-5
DOI: 10.1007/978-1-4684-0736-5

© 1973 Plenum Press, New York
Softcover reprint of the hardcover 1st edition 1973

A Division of Plenum Publishing Corporation
227 West 17th Street, New York, N.Y. 10011

United Kingdom edition published by Plenum Press, London
A Division of Plenum Publishing Company, Ltd.
Davis House (4th Floor), 8 Scrubs Lane, Harlesden, London,
NW10 6SE, England

## PREFACE

Ecological requirements are forcing a change in technology of industrially applied coatings. At present solvent-borne coatings dominate this field but in the next few years rapidly increasing use will be made of aqueous coatings, electrocoating, solvent-free liquid coatings which can be cured by U. V.-light or electronbeam, and powder coatings. This monograph describes some recent developments in these new "non-polluting" coating technologies.

It is impossible to predict how big a share each of these novel coatings will take from the conventional industrial finishes market which at present represents an annual volume of about one billion dollars. In all likelihood each will find its own specialized use. We hope that this monograph will help the reader to form a perspective of the various novel coatings, their method of application, their advantages and their limitations.

J. L. Gardon                                              Joseph W. Prane

Rahway, N. J., and Elkins Park, Pa.
October 15, 1972

# CONTENTS

# STATUS OF AIR POLLUTION CONTROL REGULATIONS

Francis Scofield

National Paint and Coatings Association
1500 Rhode Island Avenue, NW
Washington, DC 20005

On the 30th of December 1970, the President signed the Clean Air Act of 1970 which constitutes the basis of air pollution control regulations as they are being developed. This provided that the Environmental Protection Agency (EPA) should issue national ambient air standards, which was done on April 30, 1971. These provided nationwide standards for particulate matter, sulfur dioxide, carbon monoxide, oxides of nitrogen, hydrocarbons, and photochemical oxidants. The last two arising from coatings solvents are the two of concern to the coatings industry.

The Clean Air Act of 1970 required that states produce implementation plans by which the national ambient air standards would be achieved by 1975. These standards were submitted as of January 31, 1972, and on May 31 the EPA either accepted them or rejected the plans in part, with suggestions as to changes which would be necessary to achieve the ambient air standards by 1975.

Some 16 states had concentrations of hydrocarbons or photochemical oxidants exceeding the national standards. Of these, ten produced implementation plans including regulations which might be expected to achieve the standards in the prescribed time. Five states and the District of Columbia either did not produce standards satisfactory to EPA or omitted some essential requirement.

The District of Columbia, with little or no industry, could only reduce hydrocarbons and photochemical oxidants by traffic control regulations and will not be discussed further.

Indiana, Tennessee, and Louisiana have moved to adopt regulations satisfactory to the EPA. Texas believes that they can meet the requirements by controls on the storage and handling of hydrocarbons and has not attempted to control organic solvent emissions.

California with some 12 air pollution control districts, each with its own regulations, has a number of minor discrepancies which EPA felt should be corrected. This is under discussion.

At the present time it can be said that those states which have regulations controlling the emission of organic solvents had in general followed the framework of Rule 66, which was adopted by Los Angeles County in 1966. However, none of the proposed regulations follows Rule 66 exactly. There are small and sometimes significant differences. It is therefore important that the individual state regulations be examined with care since no two are exactly alike.

As a further complication, the Air Quality Act of 1970 did not provide for federal "pre-emption." That is, any state or smaller jurisdiction might adopt their own regulations if they were more stringent than the federal requirements.

New York City and Philadelphia have adopted regulations which resemble Rule 66, but have some aspects which are different and more stringent. Meanwhile, some other states even though not required by the Air Quality Act are in the process of adopting regulations covering organic solvent emissions. Notably, New York State has recently adopted Part 205 which is applicable to the New York metropolitan area. It is probable that this will supersede New York City's Local Law No. 49, but this has not yet happened.

In general, it should be remembered that the controls of these regulations are generally addressed to the users of coatings rather than the manufacturer. The manufacturer of organic coatings can in general be conducted with little or no emission of solvents into the atmosphere but the application and curing of these coatings on finished products be they factory applied, as automobiles, furniture, etc., or field applied, as residences, bridges, highways, etc., require the evaporation of large amounts of solvent and therefore fall under these regulations.

So far no state has undertaken to control the evaporation of water, and therefore water-thinned coatings tend to be exempt from these regulations. However, very few water-thinned coatings are completely devoid of organic solvents. Los Angeles and several of the states have exempted water-thinned coatings which contain less than

twenty per cent volatile material. Neither of these exemptions apply
to all states, and the state regulation should be examined carefully
to make sure what exemptions are permitted.

Finally, it should be borne in mind that this is a statement of the
current situation. Many states are still in the process of developing
new regulations or revising old ones and the situation may be expected
to change, in general, for the worse with the passage of time.

# PHYSICAL CHARACTERIZATION OF WATER DISPERSED AND SOLUBLE ACRYLIC POLYMERS

William H. Brendley, Jr.
Thomas H. Haag
Rohm and Haas Company, Research Laboratories
Norristown and McKean Roads
Spring House, Pennsylvania 19477

## INTRODUCTION

The increasingly restrictive anti-pollution legislation which regulates the amount of organic solvent that can be emitted into the atmosphere has generated renewed interest in water-based polymers for general purpose industrial coatings. More specifically, interest in aqueous acrylic polymers is high because of their inherent durability, proven exposure history, and good chemical resistance. This durability characteristic is a reflection of the fact that acrylic polymers, derived from acrylate and/or methacrylate esters, are transparent to natural sunlight and, as such, do not absorb ultraviolet radiation within these wavelengths. Their durability is well attested to by the fact that all U.S. cars are coated with either thermoplastic or thermosetting acrylic finishes. The good chemical resistance is confirmed by the usage of the acrylics as appliance finishes.

## TYPES OF AQUEOUS ACRYLIC POLYMERS

There is a tendency to associate the word aqueous with emulsion; however, aqueous acrylic polymers can be subdivided into three specific types. These three types vary significantly in physical and mechanical properties and allow the coatings chemist considerable formulation latitude.  The three types of aqueous acrylic polymers are (a) aqueous dispersions or emulsions, (b) colloidal dispersions or water "solubilized" and (c) water reducible.

## Aqueous Dispersion or Emulsion Polymers

Aqueous dispersion or emulsion polymers can be defined as discrete particles of high molecular weight polymer dispersed in an aqueous medium.  Generally, the acrylic dispersions exhibit excellent toughness, good chemical and water resistance and excellent durability.  Aqueous dispersions have been utilized in trade sales paints and building products because of these attributes.  However, their utilization in general industrial coatings has been limited by rheological and application difficulties which will be discussed in more detail shortly.

## Colloidal Dispersion or Water-Solubilized Polymers

Colloidal dispersion or water-solubilized polymers can be considered to be ultra-fine particles of molecular weight intermediate between that of aqueous dispersions and true solutions.  These polymers contain polar groups, either acidic or basic, which impart a degree of solubility.  The colloidal dispersions can be considered hybrids of dispersions and solution polymers in that they possess properties characteristic of both types (1).  The properties attained with these polymers are related to the degree of solvation of the polar groups in the polymer backbone.  Solution-like property performance can be achieved by proper solubilization via pH adjustment and/or the addition of water-miscible polar cosolvent.  The acrylic colloidal dispersions are dependent on the highly hydrated salt form of an acidic or basic group for solubility (2).  Sometimes non-ionic functional groups are also incorporated into the polymer in combination with either acidic or basic components to aid in solubilization; but they are usually not sufficiently hydrated to be used alone.  Colloidal dispersions can be subdivided into

three classes dependent on their solubility characteristics:
(1) alkali-soluble -- contains acidic groups, (2) acid-
soluble -- contains basic functionality, (3) non-ionic --
contains amide or hydroxyl groups in conjunction with either
acid or basic components.

By copolymerizing monomers containing any of the above
types, it is possible to obtain acrylic polymers which are
capable of being water solubilized.  Table I lists some of
the available monomers within each classification.

### Water Reducible Resins

Water reducible resins are copolymers in which polymerization reactions are performed in water-miscible organic solvents such as alcohols or esters.  Typically, the polar groups
must be treated to produce the more soluble salt forms to
allow water reducibility.  However, these resins are true
solution polymers and, as such, have restraints which are typical of organic solution polymers (i.e., properties and viscosity markedly dependent on molecular weight which tends to
be lower than for emulsions and colloidal dispersions).

### Table I

### Functional Monomers Useful in Preparing
### Water-Solubilized Acrylic Copolymers

| Acidic | Basic | Non-Ionic |
|---|---|---|
| Methacrylic Acid | Dimethylaminoethyl methacrylate | Methacrylamide |
| Acrylic Acid | t-Butylaminoethyl methacrylate | Acrylamide |
| Itaconic Acid | Diethylaminoethyl acrylate | Hydroxypropyl methacrylate |
| Maleic Anhydride | 2-Vinyl pyridine | Hydroxyethyl acrylate |

## COMPARISON OF THE PHYSICAL PROPERTIES
## OF AQUEOUS ACRYLIC POLYMERS

Table II summarizes a comparison of the physical properties of aqueous acrylic types.

Table II

Comparison of Physical Properties of
Aqueous Acrylic Resins

| Property | Aqueous Dispersion | Colloidal Dispersion | Water-Reducible |
|---|---|---|---|
| Appearance | Opaque Exhibits Light Scatter | Translucent Exhibits Light Scatter | Clear No Light Scatter |
| Particle Size | $\geq$ 0.1 micron | ultrafine | -- |
| Self-Crowding Capacity Constant (K) | $\approx$ 1.9 | 1.0→0 | 0 |
| Molecular Weight | 1 million | 20,000-200,000 | 20,000-50,000 |
| Viscosity | low, independent of polymer M.W. | More viscosity sensitive, somewhat dependent on M.W. | Viscosity very dependent on M.W. |

Figure 1 illustrates the solution appearance of the three acrylic types in their neat form. The water-reducible sample at the left is completely clear; it has no discrete particles and exhibits no light scatter because of complete solubilization of the polymer. By contrast, the aqueous dispersion is very milky and opaque in appearance and consists of discrete, unsolubilized particles (made up of many molecules) in suspension which exhibit self-crowding capacity. The colloidal dispersion resin (middle sample) has a translucent

Figure 1. Solution Appearance of Water-Based Acrylic Resins.

appearance because of the presence of the ultrafine, highly
swollen particles which are partially solubilized by the
presence of the polar groups.  Figure 2 is a schematic representation of these resin forms.  You will note the spherical
representation of the aqueous and colloidal dispersion polymers.  Mercurio (3) has determined self-crowding constants
(K) for the various solution forms by algebraically modifying
the Mooney equation (4) and obtaining good correlation with
experimental data.

In 1951, Mooney derived an equation for uniform polymer
spheres which has the form:

$$\ln \eta_{rel} = \frac{2.5\varphi}{1-K\varphi} \tag{1}$$

where $\eta_{rel}$ is the relative viscosity, $\varphi$ is the volume
fraction of spheres, and K is a constant identified as the
self-crowding factor of spheres on each other.  This equation
proved useful because:

(a)  it reduces to the Einstein equation (5) when $\varphi$
     is infinitely small

$$\eta_r = 1 + 2.5\varphi \tag{2}$$

(b)  considers the log of viscosity to vary linearly
     with concentration

(c)  introduces the concept of space crowding effect
     on suspended particles on each other.

Mercurio modified the Mooney equation into the form:

$$\frac{1}{\ln \eta_{rel}} = 1/BC - K/2.5 \tag{3}$$

where BC is substituted for $\varphi$ and C = concentration of
polymer (g/cc) and B is an experimental constant.

Figure 3 demonstrates the fit of the Mercurio data to
the Mooney Equation and allowed the determination of the self-
crowding constant K for the three types at 1/C > 0.

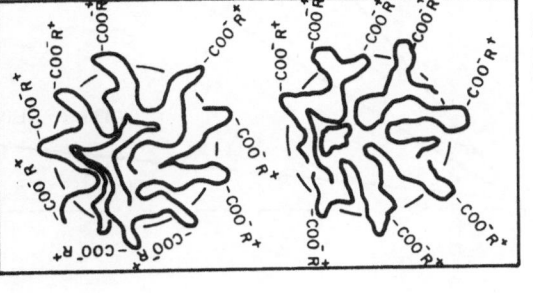
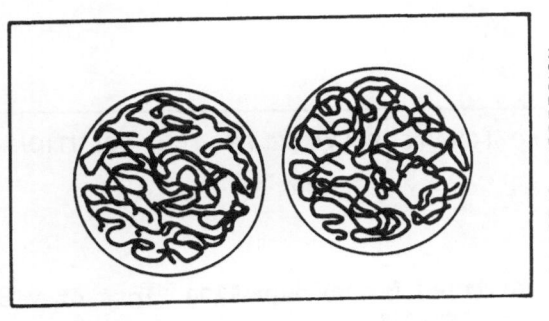

SCHEMATIC REPRESENTATION OF AQUEOUS TYPE RESINS

WATER - REDUCIBLE

COLLOIDAL DISPERSION

AQUEOUS DISPERSION

FIG. 2

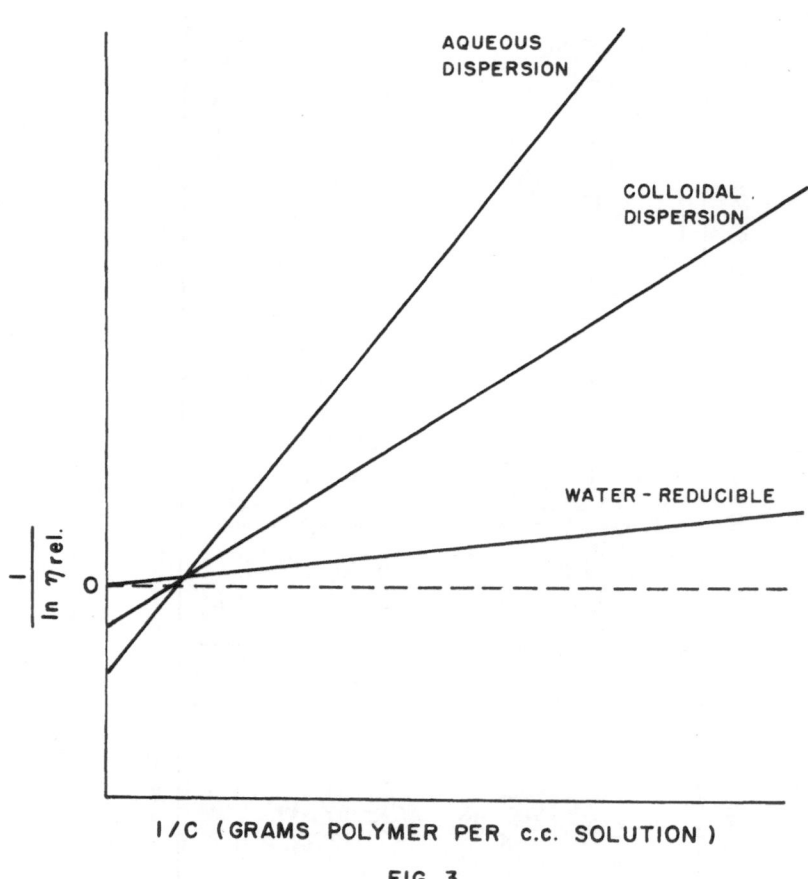

FIG. 3

Mercurio Modified Mooney Equation Plots of Viscosity versus Concentration

From these data, Mercurio postulated that an aqueous
acrylic dispersion with a self-crowding constant K value of
1.9 is quite close to the value calculated by Mooney for
simple cubic packing of rigid uniform spheres.  The  colloidal
dispersions give a value of K = 1.0 which suggests that they
too are a two phase system (they also exhibit light scatter)
and possess self-crowding capacity.  However, the colloidal
dispersions when properly solubilized approach true solution
properties even to the point of their K value approaching
zero.

## APPLICATION CHARACTERISTICS OF THE AQUEOUS TYPES

The application characteristics of the three types also
vary considerably and will influence their coating performance.
Table III summarizes these characteristics.

As previously mentioned, the aqueous dispersion resins
are polymerized to the highest molecular weight, yet exist as
discrete entities within   aqueous media.  As such, the viscosity of the solution will be relatively independent of the
polymer's molecular weight being essentially that of the
medium (in this case water).  From an inspection of Figure 4,
it becomes apparent that the aqueous dispersion will, at an
application viscosity dictated by the coating method used,
yield the highest solids content.  The colloidal dispersion
system will also yield higher application solids than the
water reducible but not as high as the aqueous dispersion
since its viscosity/solids relationship is somewhat dependent
on the polymer's molecular weight.  The water-reducible resins
are similar to typical organic solvent-based polymers with the
viscosity markedly dependent on molecular weight and consequently, yield the lowest application solids.

## RHEOLOGY CONSIDERATIONS

Besides illustrating the relative solids content of the
three aqueous types, Figure 4 can also be utilized to explain
several other phenomena associated with water based systems.

Table III

Application Characteristics of Thermoplastic
Aqueous Acrylic Polymers

| Property | Aqueous Dispersion | Colloidal Dispersion | Water Reducible |
|---|---|---|---|
| Solids at Application Viscosity | Highest | Intermediate | Lowest |
| Durability | Excellent | Excellent (-) | Very Good |
| Resistance Properties | Excellent | Good-Excellent | Fair-Good |
| Viscosity Control | Requires External Thickeners | Thickened by Addition of Cosolvent | Governed by Polymer M.W. |
| MFT Restraint | Hard Films require Coalescent | Forms Hard Films with Minimum Coalescent | None |
| Formulation Complexity | Complex | Intermediate | Simple |
| Pigment Dispersability | Poor | Good-Excellent | Excellent |
| Application* Difficulties | Many | Some | Few |
| Specular Gloss | Lowest | More Like Water Reducible | Highest |

*Ease of clean-up, foaming, method of application, substrate
 wetting rheological problems (flow, transfer).

Several application methods (roller coating, brushing) require
the coating to have good flow and leveling characteristics.
To impart good flow in a coating, the polymer in the wet state
must maintain a period of fluidity. The drying of coatings
can be effectively considered to be represented by their
solids/viscosity curves i.e., upon application, solvent is
lost and solids and viscosity increases describe the curves
in Figure 4. The film deposited from a colloidal dispersion
polymer system will arrive at an immobile state somewhat
before that of an aqueous dispersion and appreciably faster

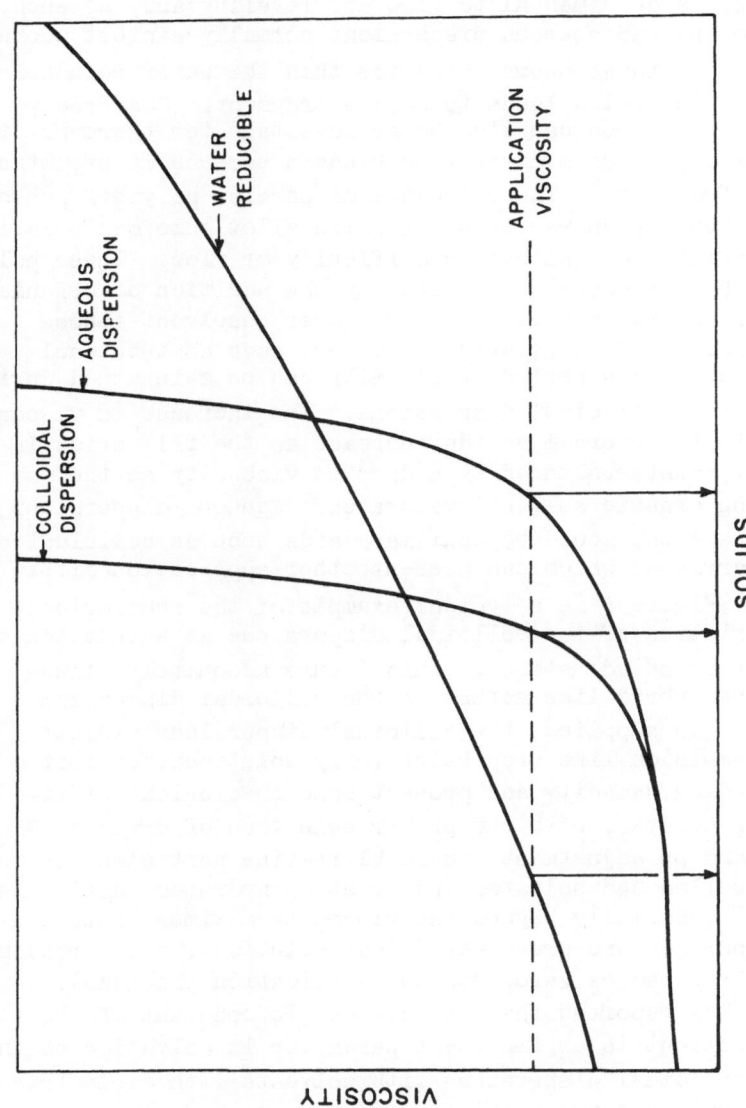

Figure 4.   Comparison of Viscosity versus Solids Relationship between the Water-Based Acrylic Resins

than that of a water-reducible system. This phenomenon, of
a rapid increase in viscosity with a relatively small increase
in solids, is detrimental to flow and leveling and, as such,
the colloidal and aqueous dispersions normally exhibit poorer
properties in these characteristics than the water reducible
polymers. This also leads to foam entrapment. Conversely,
the same phenomenon can also be an advantage for thermoplastic
colloidal and aqueous dispersion binders because it promotes
fast tackfree time. The colloidal dispersion polymers possess
some interesting characteristics which allow some manipulation
to circumvent the application difficulty of flow. These polymers can be thickened or solvated by the addition of organic
solvents. By the addition of the proper cosolvent (those
which evaporate fast relative to water, such as t-butanol
and isopropanol) a period of fluidity can be maintained during
the drying of colloidal dispersions. The increase in viscosity due to the overall solids increase as the film dries is
partially counterbalanced by a drop in viscosity as the
thickening organic solvent evaporates. Aqueous dispersions
normally are thickened by auxiliary aids such as cellulosics
or polyacrylates which can present other application difficulties. Figure 5 is a typical example of the rheological
characteristics of the colloidal dispersions as a function of
cosolvent or pH adjustment. This figure adequately illustrates the hybrid-like nature of the colloidal dispersion
polymers. As supplied, the colloidal dispersions exhibit
typical emulsion-like properties (hazy solutions, exhibit
self-crowding capacity and present some rheological difficulties); however, with the proper selection of organic cosolvent and pH adjustment, these ultra-fine particles become
highly swollen and solvated and exist as hydrated salts. In
this state, normally beyond the viscosity maximum, solution-
like properties are prevalent (clear solutions, self-crowding
constant approaches zero, minimum application problems).
Haag (6) has reported that the dielectric constant of the
organic solvent is an important parameter in solvation of the
acrylic colloidal dispersions with solvents with dielectric
constant values between 10-25 preferred.

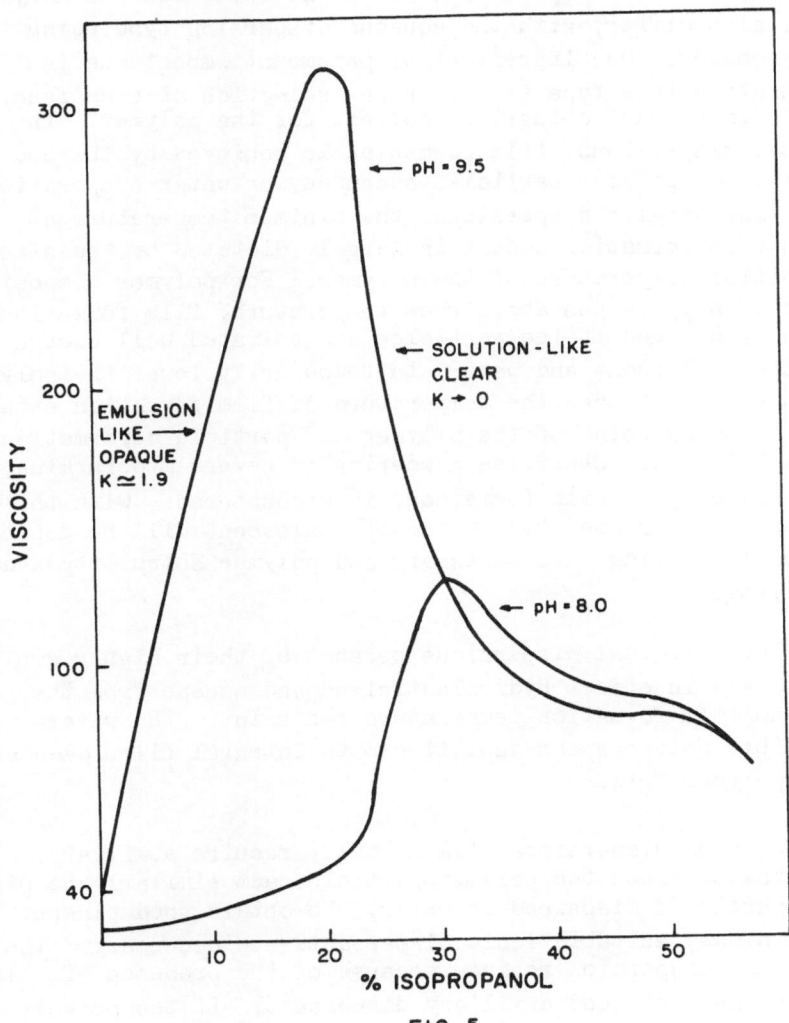

FIG. 5

Rheological Characteristics of Colloidal Dispersions
as a Function of Alcohol Cosolvent or pH

## FORMULATION COMPLEXITY

Formulation complexity between the three aqueous acrylic types also differ, with the aqueous dispersion type being the most complex. One ingredient of paramount importance in formulating this type is the proper selection of a coalescing agent, in essence a fugitive solvent for the polymer. In aqueous dispersions, film formation is achieved by the coalescence of polymer particles accompanying water evaporation (7). For acrylic dispersions, the minimum temperature at which film formation occurs is largely dictated by the glass transition temperature of the polymer. For polymer compositions with Tg values above room temperature, film formation can only proceed if the particles are solvated well enough (by the coalescent and water) to temporarily lower the polymer Tg to the point that the temperature of film formation exceeds the softening point of the polymer and particle deformation is facilitated. Otherwise powdering or severe mudcracking, evidence of poor film formation, is encountered. With the aqueous dispersions, the choice of coalescent will be dependent on the drying rate necessary and polymer solubility and stability.

The colloidal dispersions because of their highly swollen nature are in effect hydroplasticized and escape from the minimum film formation temperature restraint. The water-reducible polymers can deposit smooth integral films even at high polymer Tg's.

Aqueous dispersions will normally require auxiliary dispersants since the polymer particles act similarly to pigment particles dispersed in water. To obtain good pigment dispersions, suitable ionic dispersants are necessary. The colloidal dispersion resins, because of the presence of polar groups, may not need auxiliary dispersant. If the polymer is adequately solubilized, then the polar groups will aid in dispersing the pigment. The water-reducible resins are usually quite polar and will disperse pigment typically like organic-solvent resins without the need for extra wetting aids. The presence of some organic cosolvent in the colloidal dispersion and water reducible systems also aid in reducing foaming tendencies and surface wetting. Typical formulations of the three types illustrating the varying formulation complexity is illustrated in Table IV.

Table IV

Ball Mill Formulations

| Aqueous Dispersion | Colloidal Dispersion | Water Reducible |
|---|---|---|
| **Grind for 16 hours:** | | |
| 450 Pigment | 450 Pigment | 160 Pigment |
| 450 Emulsion | 450 Resin | 172 Resin |
| 100 H$_2$O/Cosolvent 1/1 | 100 H$_2$O/Cosolvent 9/1 | 118 H$_2$O/Cosolvent 9/1 |
| 13.5 Tamol 731 (dispersant) | 5 DMAE | 3 DMAE |
| 6.8 Triton CF-10 (wetting agent) | 2 NDW | |
| 2.0 NDW (defoamer) | | |
| **Let Down:** | | |
| 89 Paste | 89 Paste | 100 Paste |
| 60 Emulsion | 140 Resin | 57.5 Resin |
| 23 H$_2$O/Cosolvent 9/1 | 21 H$_2$O/Cosolvent 9/1 | H$_2$O/Cosolvent 1/1 |
| External Thickener | -- | -- |
| P/B = 40/60 | P/B = 40/60 | P/B = 40/60 |
| 58% Solids | 50% Solids | 44% Solids |
| 16 Sec. #4 Ford Cup | 14 Sec. #4 Ford Cup | 30 Sec. #4 Ford Cup |

SPECULAR GLOSS

Specular gloss of a pigmented coating is influenced by polymer composition, polymer molecular weight, compatibility and surface smoothness. Since aqueous dispersions have less surface smoothness (because of their particle size) and are less efficient pigment wetters than solution coatings, they

have the lowest gloss values. Typically, the water reducible
polymers yield the highest specular gloss value and the best
clarity and depth of image. The colloidal dispersions, when
properly solubilized, act like solution polymers $(K \rightarrow 0)$ and
therefore have gloss values which approach that of the water
reducible systems.


## TOUGHNESS AND DURABILITY

Assuming similar polymer compositions, the aqueous dispersions will produce the toughest films and yield the best
resistance properties although their water resistance is
sometimes compromised by the surfactants and dispersants and
soluble thickeners they contain. This is a reflection of
their much higher molecular weight and greater number of
molecular entanglements. The colloidal dispersions are
intermediate in toughness ranking and the water-reducible,
the lowest; again strictly a molecular weight relationship.
Of course, crosslinking reactions can greatly alter the
properties.

As previously mentioned, the acrylic resins offer excellent durability characteristics because of their composi-
tional nature. Nonetheless, within this excellent durability
ranking, a relative ranking of the three aqueous classes is
possible. Durability will generally relate to molecular
weight and, at similar composition, the highest molecular
weight polymer (assuming good film formation) should yield
the best durability. The colloidal dispersions will be
slightly poorer and the water reducible the least durable
within the class of excellent durability.


## OTHER MAJOR PROBLEMS OF WATER BASED RESINS
## FOR INDUSTRIAL COATINGS

Aside from the flow problem previously discussed, two
additional obstacles which limit the acceptance of water
based resins for general industrial usage are the problems
associated with substrate wetting and blistering tendencies.

## SUBSTRATE WETTING

Zisman (8) has postulated that in order to wet a substrate, the critical surface tension of the coating should be lower than the critical surface tension of the substrate. The high surface tension of water ($\approx$ 72 dynes/cm) thus presents substrate wetting problems not normally encountered with organic solvent ($\approx$ 30 dynes/cm) systems. Gardon (9) has shown both theoretically and experimentally that there is a direct relationship between the critical surface tension and the solubility parameter of the polymer. His equation states

$$\gamma_c = 3.61 \times \delta_{polymer} \tag{4}$$

where $\gamma_c$ is the theoretical surface tension of the polymer and $\delta$ is the solubility parameter derived from the method of Small (10). The critical surface tension values of several substrates and homopolymers are listed in Tables V and VI.

### Table V

### Critical Surface Tension of Solid Surfaces

| Surface | Critical Surface Tension dynes/cm |
|---|---|
| Tinplate | 35 |
| Bonderized steel | 40-45 |
| Alkyd finish | 65 |
| Glass | 70 |

One way of improving the wetting of water based systems is to lower the surface tension of the system. The addition of water-miscible cosolvents or wetting agents are sometimes beneficial.

Table VI

Critical Surface Tensions of Polymers (11)

| Surface | Critical Surface Tension dynes/cm |
|---|---|
| poly(propylene) | 29 |
| poly(ethylene) | 31 |
| poly(styrene) | 33 |
| poly(methylmethacrylate) | 38 |
| poly(methyl acrylate) | 41 |
| poly(ethyl acrylate) | 35 |
| poly(vinyl chloride) | 40 |
| poly(vinylidene chloride) | 40 |
| poly(vinylidene fluoride) | 25 |
| poly(tetrafluorethylene) | 18 |

## BLISTERING TENDENCIES

A second problem noted with aqueous coatings is a tendency toward blistering when applied to non-porous substrates and then baked. This is not totally unexpected since water has a high latent heat of vaporization ($\approx 540$ calories/g) relative to most organic solvents ($<100$ calories/ g). This problem can be minimized if sufficient air dry time is allowed prior to baking. This air dry time can range from a few minutes to one half an hour depending upon specific formulation, film thickness (thin film best), substrate type (porous best) and ambient conditions (temperature and relative humidity).

A second type of blistering encountered is the appearance of dense, small micro-blisters which can develop during the air dry period. This phenomenon is related to the flow characteristics of the polymer. Refering again to Figure 4 we see that as the aqueous systems (emulsion and colloidal dispersions) dry, there is a rapid increase in viscosity which causes a rapid set up in the film which prevents flow and traps air which during baking may break through the film and cause blistering. Utilization of organic solvents which solvate the colloidal dispersion or extend the "open time" of emulsion systems tend to alleviate this type of blistering.


## SUMMARY

In summary, this presentation has attempted to define the various types and physical characterizations of aqueous acrylic polymers. Also, the relationship of polymer coating properties to rheology and solubility parameters has been discussed, and some of the common problems of water based systems have been associated with these characteristics. These basic views will allow a more knowledgeable approach to the use of the various systems in industrial coatings.


## REFERENCES

(1) A. Mercurio, C. H. Gaiser, Resin Review, 19, p. 16 (1969).
(2) Rohm and Haas Technical Report TMM-9 (1964).
(3) A. Mercurio, Canadian Paint and Varnish, Sept. (1964).
(4) M. Mooney, J. Colloid. Sci., 6, 162 (1951).
(5) A. Einstein, Ann. Physik., 19, 289 (1906); 34, 591 (1911).
(6) T. Haag, unpublished results, Rohm and Haas Company.
(7) G. L. Brown, J. Polymer Science, 22, 423 (1956).
(8) W. Zisman, Ind. Eng. Chem., 55, No. 10 (1963).
(9) J. Gardon, reported in Rohm and Haas Technical Report TMM-9.
(10) P. A. Small, J.Appl. Chem., 3, 71 (1953).
(11) J. Swell, Mod. Plas., 66, June (1971).

# METHYLATED UREA FORMALDEHYDE CROSS-LINKING AGENTS IN AQUEOUS EMULSIONS

Leonard J. Calbo
Industrial Chemicals and Plastics Division
American Cyanamid Company

1937 West Main Street
Stamford, Connecticut

## INTRODUCTION

The trend toward tougher anti-pollution standards has had a tremendous impact on the coatings industry. In the recent past, research and development efforts have been directed toward new coatings or coating methods which can fill existing customer needs while eliminating the associated pollution problems. Much of that effort has been or is going into water based coatings. The principal advantages and disadvantages of water based coatings over conventional organic systems are tabulated below. In the past, the higher heat input and longer flash-off time generally required for water based coatings may have hindered the growth of these systems. In addition, water based systems were found to be slower curing. To overcome these drawbacks, new developments in polymer systems and cross-linking agents were needed. In the late 1960's water soluble, fast-curing, partially methylated melamine resins were introduced and found immediate acceptance in water- and solvent-based systems. With the subsequent development of methylated urea formaldehyde cross-linking agents, however, even faster curing speeds are now obtainable.

COMPARISON OF WATER AND CONVENTIONAL ORGANIC
BASED COATING SYSTEMS

| A   ADVANTAGES OF WATER | B   DISADVANTAGES |
|---|---|
| 1    FIRE RISK, INSURANCE | 1    HIGH HEATPUT, LONGER FLASH–OFF TIME |
| 2    TOXICITY, LARGE OPEN TANKS, VENTILATION | 2    SURFACE PREPARATION |
| 3    SOLVENT COST | 3    HIGH ACID CONTENT OF WATER SOLUBLE RESINS |
| 4    MULTIPLE SOLVENT STORAGE FACILITIES | 4    EMULSION SYSTEMS SUBJECT TO COAGULATION BY HIGH SHEAR AND FREEZE–THAW CYCLES |
| 5    NON–POLLUTING |  |
| 6    EASY CLEAN–UP | 5    POORER GLOSS |
| 7    SUPERIOR FILM PROPERTIES, EMULSION SYSTEMS | 6    DIFFICULT TO REPAIR DAMAGED FILMS |
| 8    UTILIZATION IN ELECTRODEPOSITION |  |

## DISCUSSION

Water based coatings include both water soluble resins and water dispersed resins, or emulsions. Either system may be thermoplastic or thermosetting. A comparison of water soluble and water dispersed systems is shown in Table I. The feasibility of preparing stable, one package thermosetting acrylic emulsion systems containing amino crosslinking agents was first demonstrated in these laboratories in 1966.[1] Because of the possibility of obtaining superior film properties, therefore, a thermosetting aqueous acrylic emulsion cross-linked with a methylated urea formaldehyde resin was chosen for study.

### Emulsion Preparation

The emulsion was prepared from a mixture of butyl acrylate, acrylonitrile, hydroxyethyl acrylate and acrylic acid, the formulation appearing in Table II. The monomer mixture was first emulsified in the presence of $NH_4OH$ and one-half the water charge and slowly fed into a flask containing the remaining water heated to 90°C. Emulsifying agent and catalyst were added to both the reactor and monomer mixture. The monomer feed required $3\frac{1}{2}$ hours, or was added at such a rate as to maintain a gentle reflue temperature of 90°C. Following reaction the emulsion was strained through a flannel cloth to remove any traces of coagulum.

TABLE *I*

## COMPARISON OF THERMOSETTING WATER SOLUBLE AND WATER DISPERSED (EMULSION) SYSTEMS

1. STABLE THERMOSETTING SYSTEMS CAN BE PREPARED FROM BOTH WATER SOLUBLE AND EMULSION POLYMERS USING METHOXYMETHYL MELAMINE AND UREA CROSS–LINKING AGENTS.

2. WATER SOLUBLE POLYMERS ARE USUALLY EASIER TO FORMULATE.

3. PRESENCE OF EMULSIFIERS AND WETTING AGENTS IN AN EMULSION PAINT MAY EFFECT THE WATER RESISTANCE AND CAUSE FOAMING.

4. EMULSION SYSTEMS PERMIT THE USE OF HIGH SOLIDS ENAMELS AT APPLICATION VISCOSITY, AND THE USE OF HIGH MOLECULAR WEIGHT POLYMERS.

5. EMULSION SYSTEMS YIELD FILMS WITH A SUPERIOR COM–BINATION OF MECHANICAL PROPERTIES.

6. EMULSION COATINGS HAVE BETTER ALKALI RESISTANCE BECAUSE THEY CONTAIN FEWER FREE CARBOXYL GROUPS.

TABLE *II*

### EMULSION FORMULATION

|  | WT. | WT. % | MOLES | MOL % |
|---|---|---|---|---|
| ACRYLONITRILE | 1512 | 42 | 26.5 | 60.9 |
| BUTYL ACRYLATE | 1656 | 46 | 12.9 | 29.7 |
| $\beta$–HYDROXYETHYL ACRYLATE | 360 | 10 | 3.1 | 7.1 |
| ACRYLIC ACID | 72 | 2 | 1.0 | 2.3 |
| WATER (DEIONIZED) | 4210 | | | |

CATALYST: POTASSIUM PERSULFATE, 0.28% ON MONOMER MIX
EMULSIFYING AGENT: DUPONOL C, 2.9% ON MONOMER MIX

### EMULSION PROPERTIES

| | |
|---|---|
| SOLIDS | 43% |
| COAGULUM | NEGLIGIBLE |
| VISCOSITY | 28.3 POISES AT 10 RPM |
| BROOKFIELD 25°C. | 6.2 POISES AT 100 RPM |
| PARTICLE SIZE | 300–1000 ANGSTROMS |
| LATEX pH | 5.8 |

## Amino Resin

The methylated urea formaldehyde cross-linking agent used in this study is a water soluble, partially methylated urea formaldehyde reaction product of low molecular weight. It was previously demonstrated in our laboratories that exceptionally fast cures could be obtained with short oil alkyds when BEETLE® 60* methylated urea resin was substituted for a butylated resin on an equal weight basis. Curing speeds 4 to 5 times faster than those of conventional systems were obtained. It was thought possible that these fast curing properties could be incorporated into water based systems.

The reactive functional groups present in BEETLE 60 methylated urea resin include $-NH$, $-NCH_2OH$ and $-NCH_2OCH_3$. A summary of the typical self-condensation reactions which may occur in aqueous media is illustrated in Figure 1. The tendency to self-condense, however, is greatly reduced as the degree of alkylation is increased. Consequently, the resin chemist is forced to compromise and must seek the optimum balance between stability and curing speed.

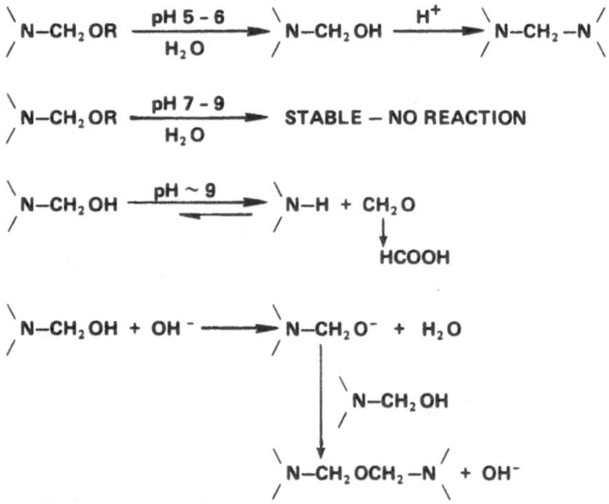

Figure 1.   Typical self-condensation reactions of amino cross-linking agents.

*   A product of American Cyanamid Co., Industrial Chemicals and Plastics Division, Wayne, New Jersey  07470

## Cure Study

The clear coating formulations studied were prepared by first neutralizing the carboxyl groups in the acrylic emulsion with ammonium hydroxide and then adding the acid catalyst. When para-toluene sulfonic acid was used as catalyst, it was first neutralized with ammonium hydroxide. The amino resin was then added to the emulsion and the solids level adjusted by the addition of water. All data were obtained on glass draw-downs baked in a forced draft oven. Similar results were achieved for a spray application on wood.

In Figure 2 it can be seen that excellent cure response can be obtained at 200°F. with the amino level as low as 16%. However, faster cures and better properties overall can be obtained at an amino level of 28.5%. When the amino level is raised to 50% a plasticizing phenomenon occurs with the excess cross-linking agent helping initially

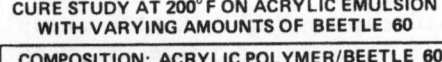

. FIGURE 2

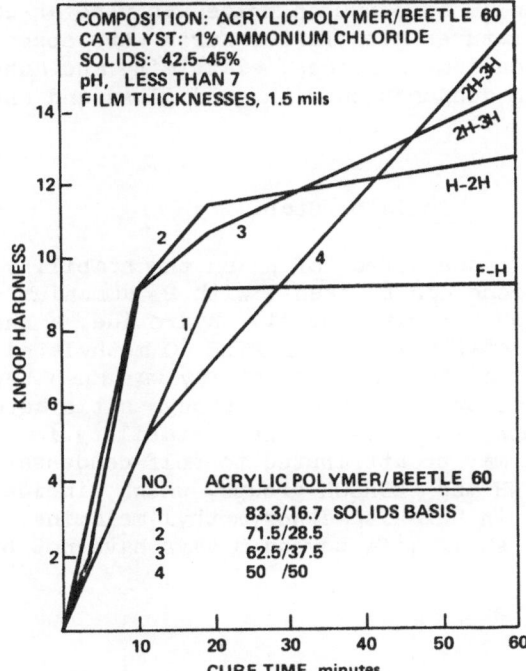

CURE STUDY AT 200°F ON ACRYLIC EMULSION WITH VARYING AMOUNTS OF BEETLE 60

COMPOSITION: ACRYLIC POLYMER/BEETLE 60
CATALYST: 1% AMMONIUM CHLORIDE
SOLIDS: 42.5–45%
pH, LESS THAN 7
FILM THICKNESSES, 1.5 mils

| NO. | ACRYLIC POLYMER/ BEETLE 60 |
|-----|----------------------------|
| 1 | 83.3/16.7 SOLIDS BASIS |
| 2 | 71.5/28.5 |
| 3 | 62.5/37.5 |
| 4 | 50 /50 |

to plasticize the film thus making it softer. Undoubtedly
some self-condensation of the amino resin is occurring since
there is only a limited number of hydroxyl groups available
for cross-linking. Ordinarily one would expect self-condensation to lead to harder films, and eventually it does
as shown by the Knoop hardness after 60 minutes. However,
since we are dealing with such low molecular weight cross-
linking agents ($< 500$), initial self-condensation will en-
hance plasticization.

## Comparison with Hexakis[methoxymethyl]melamine

When compared to CYMEL® 300A, much faster cures were
observed for the methylated urea resin. At 175°F. there
is a dramatic difference, with BEETLE 60 methylated urea
resin achieving a Knoop hardness greater than 6 after a 30
min. bake (Figure 3). The melamine resin showed very little
response at this schedule. Even after 60 min. the melamine
resin was still soft and tacky whereas the film cured with
the methylated urea resin had risen to a Knoop hardness of
8 and a pencil hardness of H-2H. Similar results can be seen
in Figure 4 when 1% ammonium chloride was used as catalyst.
At higher temperatures even faster cures were observed
(Figure 5). Excellent hardness, elasticity and adhesion
were obtained in conjunction with good gloss and chemical
resistance.

## Emulsion Stability

In Table III the effect of pH on the stability of the
catalyzed emulsions can be seen. With 2% ammonium chloride
neutralized to pH 8.3 with ammonium hydroxide, a shelf life
of 4-5 days is obtainable with BEETLE 60 methylated urea
resin. As expected the shelf stability was much better with
hexakis[methoxymethyl]melamine. Although satisfactory for
most applications, the limited shelf stability for the
methylated urea may be attributed to self-condensation of
the available -NH and -NCH$_2$OH groups, which, incidentally,
are not present in hexakis[methoxymethyl]melamine. Efforts
to increase the shelf life beyond 6 days have not been too
successful.

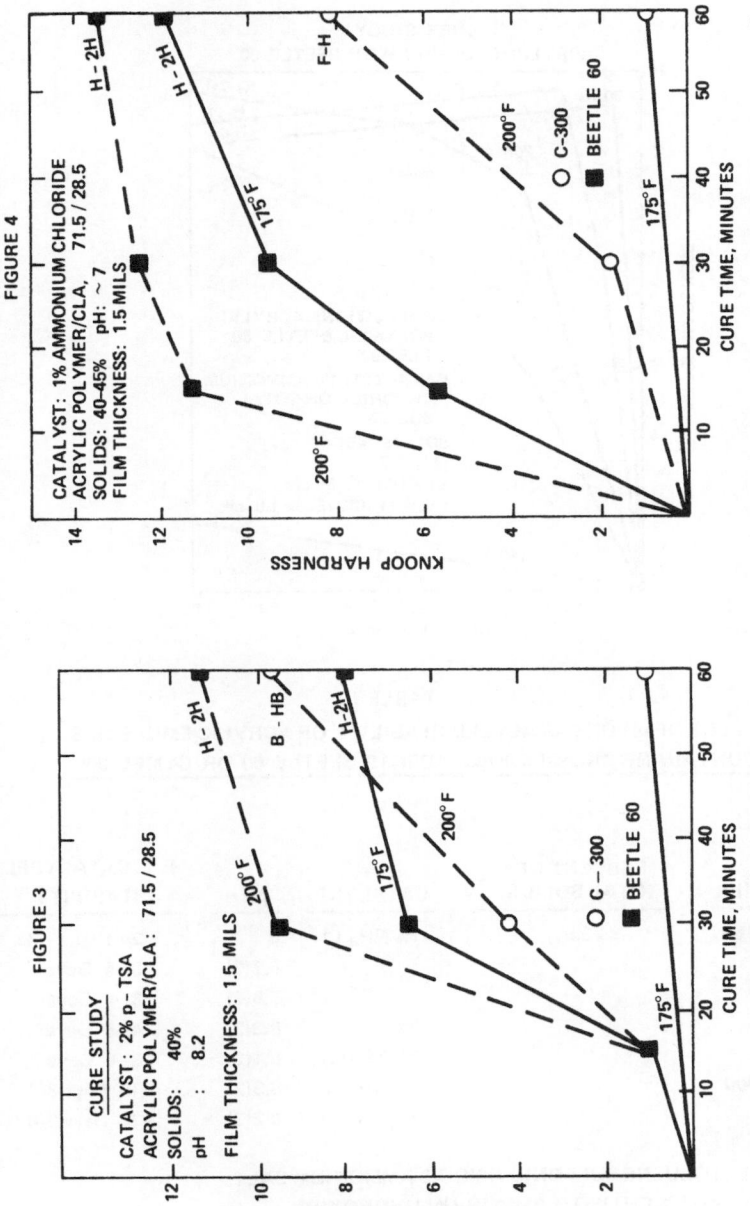

Comparison of BEETLE 60 methylated urea resin with CYMEL 300 hexakis[methoxymethyl] melamine using 2% p-toluene sulfonic acid (Figure 3) and 1% ammonium chloride (Figure 4) as catalyst.

FIGURE 5

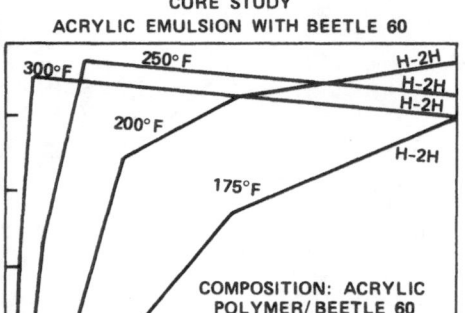

CURE STUDY
ACRYLIC EMULSION WITH BEETLE 60

COMPOSITION: ACRYLIC
POLYMER/ BEETLE 60
71.5/28.5
CATALYST: 1% AMMONIUM
CHLORIDE ON TOTAL
SOLIDS
SOLIDS: 45%
pH, 6.7
VISCOSITY, 85 cps
FILM THICKNESS, 1.5 mils

TABLE *III*

EFFECT OF pH ON CATALYZED STABILITY OF ACRYLIC EMULSIONS
CONTAINING CROSSLINKING AGENTS BEETLE 60 OR CYMEL 300

| MODIFIER | PERCENT OF TOTAL SOLIDS | CATALYST | pH | R.T. CATALYZED STABILITY |
|---|---|---|---|---|
| BEETLE 60 | 28.5 | 1% $NH_4$ Cl | <6 | <24 Hrs. |
| " | " | " | 6.7(2) | 3—4 Days |
| " | " | " | 7.8(2) | 3—4 Days |
| " | " | 2% " | 8.3(2) | 4—5 Days |
| " | " | 2% TSA(1) | 8.1(2) | 3—6 Days |
| CYMEL 300 | " | 1% $NH_4$Cl | 8.3(2) | >15 Days 3/4 |
| " | " | 2% TSA(1) | 8.2(2) | >14 Days 3/4 |

(1)  TOLUENE SULFONIC ACID AS AMMONIUM SALT
(2)  ADJUSTED WITH AMMONIUM HYDROXIDE

## Cross-linking Mechanism

H. Petersen[2,3] and S. Vail[4] have proposed a reaction mechanism to describe the acid catalyzed cross-linking of cellulosic derivatives with alkoxymethyl and hydroxymethyl urea derivatives. The same relationships should apply in this case since cross-linking is occurring between the hydroxyl group of the acrylic polymer and alkoxymethyl groups of the urea moiety. As depicted in Figure 6, initial protonation occurs at the ethereal oxygen. Equilibrium is rapidly established and the collapse of the resulting inter- mediate can lead to a carbonium ion or its resonance hybrid, an immonium ion, which then reacts with the hydroxyl group of the acrylic polymer to yield a C-O-C cross-link. It should be borne in mind, however, that self-condensation of the urea resin is also occurring, which undoubtedly is con- tributing to the faster curing rates observed.

FIGURE 6

POSSIBLE
CURING MECHANISM

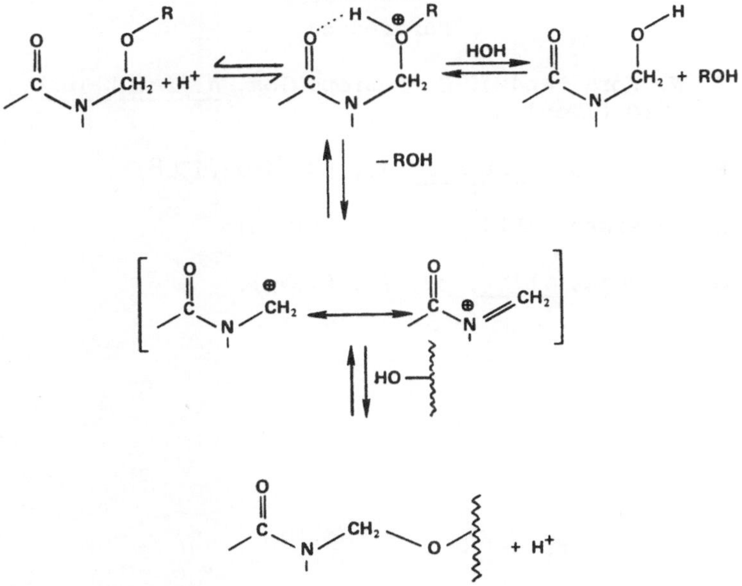

R = CH₃ or H

## SUMMARY

We have demonstrated that faster curing speeds and lower cross-linking temperatures are now possible in aqueous emulsion systems.  Excellent cures can be obtained at temperatures as low as 175°F.  The stability of the uncatalyzed emulsion is approximately 2 weeks while that for the catalyzed emulsion is 4-5 days.  Future efforts will concentrate on achieving greater shelf stability without any sacrifice in curing speed.  It may be possible that the recently developed 100% solids BEETLE 65 methylated urea cross-linking agent can fill this need.

## Acknowledgement

The author wishes to especially thank Mr. Lennart Lundberg for preparing the emulsions and performing the cure study, and would also like to thank Mr. Werner Blank for many helpful discussions during the preparation of the manuscript.

## References

(1)  J. N. Koral and J. C. Petropoulos, J. Paint Tech., 38, 610 (1966).

(2)  H. Petersen, Text. Res. J., 38, 156 (1968).

(3)  H. Petersen, ibid., 40, 335 (1970).

(4)  S. L. Vail, ibid., 39, 774 (1969).

# POLYMER PARAMETERS AFFECTING AQUEOUS TOPCOAT SYSTEMS

Robert A. Ottaviani

General Motors Research Laboratories

Warren, Michigan 48090

## INTRODUCTION

Of major concern in the field of industrial surface
coatings is the elimination or reduction of volatile organic
solvents in order to comply with existing or proposed pollution control laws.  The substitution of water for the
organic solvents or its use as the continuous phase of a
coating system is one manner of approaching this problem.
However, it must be stressed that the appearance, ease of
application, durability, and the myriad of other desirable
properties of our current coating systems should not be
compromised in order to achieve a pollution free system.

The initial investigations carried out in the area of
water soluble and water dispersible coatings in the 1950's
raised many obstacles against their use as automotive
topcoats.[1]  It is the purpose of this investigation to preform
an overall study of these water systems and the factors
influencing their use.  The intent was to determine the
feasibility of using such systems in light of the advances
that have occurred in the field of science and technology
within the coating industry in the past twenty years.

## EXPERIMENTAL

The acrylics and styrene modified acrylics were chosen
as the vehicle because of the numerous desirable properties
they exhibit as a class.  In addition, their compositions
were patterned after current organic-based automotive
finishes.


### Emulsion Polymerization

In a typical experiment a 12 oz. pop bottle, rinsed
prior to use with distilled water, a dilute (0.6 g/1000 ml)
aqueous solution of GAFAC RE-610 emulsifier, and distilled
water, was charged with 138 g of deaerated distilled water
(briefly bubbled with nitrogen, or boiled and cooled under
nitrogen flow), 1.8 g of GAFAC RE-610 emulsifier, 89.6 g of
methyl methacrylate (MMA), and 22.4 g of n-butyl acrylate
(n-BA) under a nitrogen atmosphere.  The bottles were always
charged in the above order and then capped with a Neoprene
septum and perforated crown cap.  The bottle was tumbled in
a water bath at 65°C for two hours, and one ml (0.0125 g)
of an aqueous potassium persulfate solution was added with
a syringe.  The bottle was agitated at 65°C for an additional
16 to 24 hours.  The resulting emulsion was bluish in color
with a pH of 3.0 ±.5.  The polymer latex, so produced, was
used as is or collected by precipitation or steam distillation
and filtered, washed and dried.  The more successful recipes
were scaled up to produce sufficient quantities for further
study by carrying out the emulsion polymerization in one or
two liter resin kettles.

Post-emulsification of the above materials or of
materials obtained via solution polymerization (see below)
was carried out using a high shear Cowles mixer or a Waring
blender with a high shear head attachment.


### Solution Polymerization

In a typical experiment, 300 ml of dry toluene (stirred
over sodium for 24 hours and distilled from sodium prior to
use) was added to a one liter four-necked resin kettle
equipped with stirrer, thermometer, condenser, and addition
funnel with gas inlet tube.  To the addition funnel was added
200 ml of dry toluene, 70.1 g methyl methacrylate, 99.5 g of

n-butyl methacrylate, 20.7 g of methacrylic acid, 51.8 g of 2-hydroxypropyl methacrylate, and 1.64 g of 2,2'-azobis[2-methylpropionitrile]. All monomers were distilled immediately before use. Nitrogen was bubbled through the solution in the funnel and in the reaction kettle for 0.5 hour to remove oxygen. The reaction vessel was heated to 80°C, the stirrer started, and the monomer-initiator solution added dropwise over a two hour period. The temperature rose to 100-110°C during the addition and was maintained at this temperature for 4-5 hours. The polymeric material was collected by precipitation from hexane, filtered, washed, and dried in the usual manner. This material could then be dissolved or dispersed in water using ammonia or an organic amine.

## Paint Preparation

In a water-soluble type formulation, the pigment (in most cases 99.7 g $TiO_2$, 0.05 g carbon black, and 0.25 g ferrite yellow) was ground on a shaker mill using glass beads or with a batch sand mill. When possible, the pigment grinding was carried out in a minimum amount of water in order to reduce foaming. Since the water-soluble vehicles contained up to 20% of exempt organic solvents, these were usually employed to aid dispersion.

In an emulsion type formulation, water, dispersant, pigment and cellosolve were added in this order and dispersed with a Cowles Dissolver. Incremental amounts of the emulsion vehicle (30% N.V.) was slowly added, and dispersion continued briefly in order to minimize foaming. The remainder of the emulsion was then added and stirred. The large scale use of additives was avoided as much as possible.

## Panel Preparation

Phosphated cold rolled steel panels were used as the substrate in those experiments employing electrodeposition primers. After baking at 350°F for 15 minutes, a commercial primer-surfacer was applied and baked for 30 minutes at 350°F. The left half of a number of the panels were wet sanded with 400 grit sandpaper, rinsed with distilled water, and given a dry off bake at 300°F for 10 minutes. The topcoat was applied and baked using a graduated bake

schedule (room temperature to 200°F to 325°F).  In those
incidents employing an adhesion promoter between the
electrodeposition primer and topcoat, the topcoat was
applied wet-on-wet and then baked as above.  Phosphated cold
rolled steel panels were used in the experiments employing
primer-surfacer only.  After an initial bake of 325°F for
45 minutes, the panels were wet sanded as above, rinsed with
distilled water, and given a dry off bake of 10 minutes at
300°F.  The topcoat was applied and baked as before.

## Panel Evaluation

The following tests[2] were performed on the prepared
panels:  humidity, tape adhesion, crack cycle, condensing
humidity, gravelometer, surface distortion, chemical spotting,
and Florida exposure.

## Application Methods

Water base topcoats are of particular interest in that
application methods are generally similar to existing coating
processes.  Therefore, most application methods were limited
to use of the standard air atomization gun.  The following
topcoating technique was used in a typical coating application:

1.  Atomization pressure was maintained at 90 psi with
    a traverse speed of 2.2-3.0 ft./sec., and a
    distance of 10-14 inches between gun and panel.
2.  The coating was applied in three double passes
    with a 5-15 minute flash between coats.
3.  20-60 minute air dry prior to bake.

The use of hot spray techniques were also investigated.  The
temperature was varied between 50-85°C in these instances.

## Bake Conditions

A graduated heating schedule was used in all cases in
order to minimize solvent popping which is one of the most
difficult problems (see Results and Discussion).  The following schedule was used:

1.  Initial bake from room temperature to 200°F and
    held at 200°F for 10-15 minutes.

2.  Temperature was then raised to 325-350°F and
    held for an additional 20 minutes.

A closed-loop vapor reflow system was also investigated.
Heating was carried out in a number of chlorinated solvents
and water vapor (steam).

## RESULTS AND DISCUSSION

Since emulsion coatings yield thermoplastic films, they
are understandably attractive; and they were therefore, the
first to be investigated.  However, the emulsion system
presents numerous problems.  Formulation is difficult.
Often it required that the pigment be ground separately in
order to minimize foaming.  Stability of the system in this
step, as in subsequent steps, is difficult to control.
Coalescence of the resin under shearing forces occurs
readily.  For gloss, it is necessary to get good pigment
wetting and these systems tend not to completely wet the
pigment, thereby resulting in low gloss films.  Gloss
readings on a 20° meter gave values between 44 to 63 points.
Complete coalescence of the emulsion particles to yield a
crack free and pinhole free film is desirable.  However, it
is difficult to get complete cohesive film formation.  The
numerous additives incorporated into the emulsion systems
in order to overcome these problems usually introduce more
problems than they solve.  These additives are themselves
water soluble, and therefore, they become sites of attack
lowering the water resistance of the coating.  Such additives
were avoided as much as possible.  Because of the above
limitations, the emulsion systems were not considered suitable for a high gloss automotive finish.

In order to overcome the above problems, attention was
directed toward water soluble thermosetting vehicles.  All
of the vehicles considered here contained the carboxyl
functionality as pendant groups on the chain backbone in
order to effect solubilization via a neutralization reaction
with base and to act as

$$\text{\textasciitilde}CO_2H + B: \longrightarrow \text{\textasciitilde}CO_2^- + BH^+ \qquad (1)$$

subsequent crosslinking sites.[3]  Hydroxyl groups were also
incorporated into the polymer to aid solubilization and to
act as additional crosslinking sites.  The compositions of

the polymers were modeled after current organic-based
automotive systems.

In order to aid water solubility and keep the solution
viscosity to a minimum, the molecular weights were kept
below 25,000.[4]  Calculations of the molecular weight were
obtained from the equation[5]

$$1/\overline{DP} = 1/\overline{DP_0} + C_X (X)/(M) \tag{2}$$

where $1/\overline{DP_0}$ is the average degree of polymerization in the
absence of X, $C_X$ is chain transfer constant for X, X is the
chain transfer agent, and M is the monomer.  Since it was
observed that the presence of high molecular weight tails
in the polymer resulted in greater ease and extent of crater
and fish-eye formation, the monomers and initiator were
added dropwise to the reaction mixture.  Although this is
an an approximate method of maintaining control of the
molecular weight distribution, it was found to effectively
minimize the quantity of high molecular weight material.
Theoretically, the molecular weight distribution can be
narrowed by maintaining the instantaneous $\overline{DP}$ constant
throughout the polymerization or by maintaining the instantaneous kinetic chain length $(\upsilon_k \sim R_p/R_t)$ constant.[6]

The glass transition temperature (Tg) is another
important parameter of the vehicle that must be considered.
The materials considered here had Tg's between 30-70°C.  For
organic-based systems, it has been reported[7] that the optimum
Tg for a suitable thermoset automotive finish is between
35 and 60°C.  An approximate calculation of the Tg of the
copolymer can be obtained from the equation[8]

$$1/Tg = \sum_{n=1}^{\infty} W_n/(Tg)_n \tag{3}$$

where Tg is the glass transition temperature (K°) of the
copolymer, $(Tg)_n$ is the glass transition temperature (K°)
of the homopolymer, and $W_n$ is the weight % of a particular
monomer in the copolymer.

The crosslinking agent employed in the experimental
resins was hexamethoxymethyl melamine (HMM).  This material
undergoes effective crosslinking reactions when heated under
acid conditions.  The amine salt of the carboxyl group which

$$\text{Polymer}-\text{OH} + \begin{array}{c} CH_3OCH_2 \\ CH_3OCH_2 \end{array}\!\!N\!\!-\!\!\left[\begin{array}{c} N \\ N \quad N \end{array}\right]\!\!-\!\!N(CH_2OCH_3)_2 \quad \xrightarrow[H^+]{\Delta}$$

$$(CH_2OCH_3)_2$$

(HMM)

Polymer
$$|$$
$$O$$
$$|$$
$$CH_2$$

$$\text{Polymer}-\text{O-CH}_2-N\!\!-\!\!\left[\begin{array}{c} N \\ N \quad N \end{array}\right]\!\!-\!\!N\text{-CH}_2OCH_3$$

$$CH_2$$
$$CH_3O$$

$$(CH_2OCH_3)_2$$

+2MeOH

was used to effect solubility was the acid catalyst, although an additional acid may be used to further accelerate cure.

The acid content of the polymer was varied between 10-25 mole %. The hydroxyl content was varied between 15-30 mole %. HMM was added up to 25% by weight based on the vehicle weight above.

Since low temperature cure is of considerable interest, possibilities in this area were also investigated. It is possible to achieve curing down to room temperature through the use of ionic or coordination bonding. However, films cured in this manner were found to weather badly. Electron probe data revealed considerable random segregation of the metal complexing agent throughout the film.

The major problems encountered with all water soluble coatings were blistering, cratering, fish-eyeing and changes in the spraying conditions found to occur from day to day. These problems were due mainly to the high surface tension and high heat of vaporization of water, and to the fixed boiling point of the system as a whole.

A minimization of crater and fish-eye formation and an increase in the initial gloss of the water soluble coatings was achieved through increased vehicle wetting and use of suitable surfactants and plasticizers in their paint formulation. The blistering and day to day application variations, however, still remain as major problems. Because of the sensitivity these systems have shown to atmospheric conditions, humidity control is essential. Changes of only 10% in the relative humidity can cause drastic changes in the application characteristics of these systems and in the drying rate. Blistering can be traced to the entrapment of air and water.

## ACKNOWLEDGEMENTS

The author wishes to acknowledge A. C. Ottolini, T. P. Schreiber and E. L. White, Chemistry Department, GM Research Laboratories, for obtaining the analytical data. The technical assistance of B. J. Vannoy is gratefully acknowledged.

## REFERENCES

1.  Martens, C. R., "Emulsion and Water-Soluble Paints and Coatings," Rheinhold Publishing Corporation, Chapman and Hall, Ltd., London, 1964, Chapter 15.

2.  "Test Procedures and Specifications for Coating Materials," Coating Materials Department, Manufacturing Development, General Motors Corporation.

    ASTM Method D-2246.

    SAE Method J-400.

    ASTM Method D-2247.

    ASTM Method D-714.

3.  Saxon, R., and Lestienne, F. C., J. Appl. Polymer Sci., 8, 475 (1964).

4.  Martens, C. R., "Emulsion and Water-Soluble Paints and Coatings," Rheinhold Publishing Corporation, Chapman and Hall, Ltd., London, 1964, Chapter 2.

5.  Gregg, R. A., and Mayo, F. R., Discussions Faraday Soc., 2, 328 (1947).

    Brandrup, J., and Immergut, E. H., "Polymer Handbook," Interscience Publishers, New York, 1967, p. II-77.

6.  Osakada, K., and Fan, L. T., J. Appl. Polym. Sci., 14, 3065 (1970).

7.  South Africa Patent 495662, 1965.

8.  Klein, D. H., J. Paint Technology, 42, 335 (1970).

# PERFORMANCE CHARACTERISTICS OF A FAST CURING ACRYLIC RESIN IN ELECTROCOATING PAINTS

Girish G. Parekh
Industrial Chemicals & Plastics Division
American Cyanamid Company

1937 West Main Street
Stamford, Connecticut

## INTRODUCTION

During the last decade the electrocoating process has gained increasingly greater importance in the application of industrial paints. More recently, considerable interest has developed in an electrocoating system for white single coats requiring either a lower baking temperature or a shorter baking cycle. At present, most of the existing saturated anionic resin systems require a baking temperature of 150°C. or higher for a period of 15 to 30 minutes. The advantages of a lower baking temperature or shorter baking cycle are as follows:

a. Faster line speed resulting in a higher production rate.
b. Coating of an assembled article with plastic parts.
c. Coating of temperature-sensitive metals.
d. Lower consumption of energy.

Previous work in this laboratory dealt with the utilization of amino cross-linking agents[1] and the advantages of a saturated anionic acrylic resin[2] as the vehicle in the electrocoating systems. The present work deals with the utilization of a newly developed anionic acrylic resin XC-4015 in combination with the amino cross-linking agents for electrocoating. The aim was to have a system which would require a low baking temperature or a short baking cycle at elevated temperatures and also give non-yellowing enamels.


## EXPERIMENTAL PROCEDURE


### Electrocoating Vehicles


XC-4015 Resin.* A saturated anionic acrylic resin, commercially available as a 75% solution in n-butanol. It has acid and hydroxyl values of 70 and 60, respectively.

XC-4010 Resin.* A saturated anionic acrylic resin, commercially available as a 75% solution in 2-ethoxy ethanol. It has an acid value of 110.

CYMEL® 1116* Cross-Linking Agent. A fully etherified alkoxymethyl melamine derivative, widely utilized in electro-coating.


### Paint Preparation and Electrocoating Procedure

White pigmented formulations were prepared by blending the required amounts of acrylic resin, CYMEL 1116 amino cross-linking agent, diisopropanolamine, and UNITANE® OR-600** titanium dioxide rutile pigment, and subsequently grinding the paste on a three-roll mill. The resulting paste was stirred under high speed agitation and small portions of deionized water were added. After the phase inversion, characterized by a sudden drop in viscosity, the remaining deionized water was added in one portion to bring the paint solids to 10%. The aqueous paint bath was allow-ed to age at ambient temperature with constant stirring for at least 24 hours prior to testing. In all cases where the bath composition is not mentioned a resin/cross-linking

*Products of American Cyanamid Company, Industrial Chemi-cals & Plastics Division, Wayne, N.J. 07470

**Product of American Cyanamid Company, Pigments Division, Bound Brook, N.J. 08805

agent ratio of 75/25 and a pigment/binder ratio of 40/100 was used. The amount of diisopropanolamine used corresponded to 45% neutralization of the available carboxylic acid groups. In this study the resin/cross-linking agent ratio of 75/25 was found to be the optimum level for good film properties. Generally, alkanolamines were better solubilizing agents than alkylamines; however, the use of very high boiling alkanolamines such as triethanolamine or triisopropanolamine was not practical. Traces of high-boiling alkanolamines remaining in the electrodeposited film retarded the cure.

After electrocoating, the panels were rinsed with deionized water. For most of the evaluations the coating was performed on cold rolled steel panels with Bonderite 37 conversion coating (Parker Rustproof Company, Detroit, Michigan). For electrocoating, a D.C. power supply model 500 AB (Ransburg Electrocoating Corp., Indianapolis, Indiana) was used at 150 V to 250 V for 60 to 90 seconds.

## Cure Cycle

The Bonderite 37 pretreated cold rolled steel panels were electrocoated in baths containing XC-4010 and XC-4015 resins as vehicles. In both baths the resin-to-cross-linking agent ratio, pigment-to-binder ratio, and the amine levels were the same. After rinsing with deionized water, the electrocoated panels were baked at 125°C., 150°C., 175°C., and 200°C. for periods ranging from 2 minutes to 60 minutes. In another study electrocoated panels obtained from the XC-4015 bath were cured for 20 minutes at temperatures ranging from 110°C. to 180°C. In all cases, the film thickness was 0.8 mil.

## Degree of Cross-Linking

The degree of cross-linking of the cured film was measured in terms of Knoop hardness (25 gm. load) obtained with Tukon microhardness tester (Wilson Instrument Div., American Chain and Cable Company, Inc., New York, N.Y.). This was a good practical method for measuring the degree of cross-linking since it was found that the higher the Knoop hardness of the film the greater the resistance to organic solvents, especially esters and ketones.

## Bath Stability

The electrocoating bath with XC-4015 resin, having a
resin/cross-linking agent ratio of 75/25 and diisopropanolamine neutralization of 60% of the available carboxyl
groups, was aged with constant stirring over a period of
7-8 weeks at ambient temperature. At weekly intervals the
bath was tested for pH and conductivity. Cold rolled steel
panels pretreated with Bonderite 37 conversion coating and
aluminum panels pretreated with Alodine 1200S conversion
coating (Amchem Corporation, Ambler, Penna.) were electrocoated and cured at 125°C. for 20 minutes. The gloss of
the film at 0.8 mil thickness was measured at 60° and 20°.
All bath measurements such as pH and conductivity were
carried out at ambient temperature. The water level was
adjusted prior to coating in order to compensate for evaporation losses. No organic solvent was added during aging.

## Throwing Power Tests

The throwing power tests were conducted in a cylindrical tank (15" x 6.5" dia.). As shown in Figure 1, two cold
rolled panels (4" x 12") with Bonderite 37 conversion coating were inserted in the tank filled with XC-4015 based
paint. Both panels were kept 1/8" apart and parallel to
each other on a non-conductive stand. The anode was connected to the two panels and the cathode to the wall of the
tank. The tests were performed at different voltages and
amine levels.

## EXPERIMENTAL RESULTS

### Comparison of Cure Rate of Electrodeposited
### Films Based on XC-4010 and XC-4015

The curing results of XC-4010 and XC-4015 resins are
shown in Figures 2 and 3. In combination with CYMEL 1116
cross-linking agent, the XC-4015 resin requires a temperature for cross-linking at least 25°C. lower than that required by the XC-4010 resin. The minimum cure temperature
required for XC-4015 was 125°C. while XC-4010 required at
least 150°C. At 125°C. cure temperature the film obtained
from the XC-4010 system was soft, tacky and uncured. At
elevated temperatures, XC-4015 cross-links at much faster

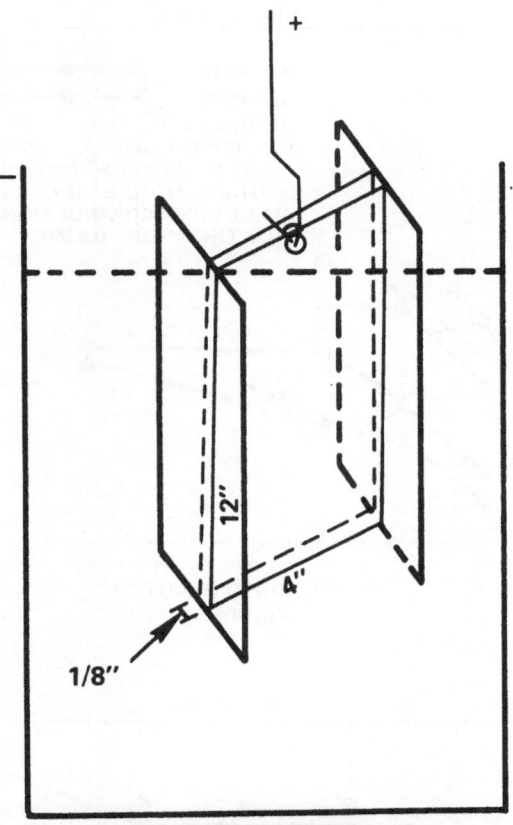

**THROWING POWER TEST**

**FIGURE 1**

rate than XC-4010.  For comparable cure at 175°C., XC-4010 requires nearly 15 minutes while XC-4015 requires only 5 minutes.  For complete cure at 200°C., XC-4010 and XC-4015 require 10 minutes and 3 minutes, respectively.  It was also found that the XC-4015 with a film thickness of 0.2-0.4 mil could be fully cured at 225°C. within 60 seconds.  In our cure study of XC-4015 resin with other amino cross-linking agents marketed as CYMEL 1123 and CYMEL 1132, XC-4015 did not cure either at a lower temperature or at a faster rate at higher temperature.  The cure rate was found to be comparable to that of XC-4010.

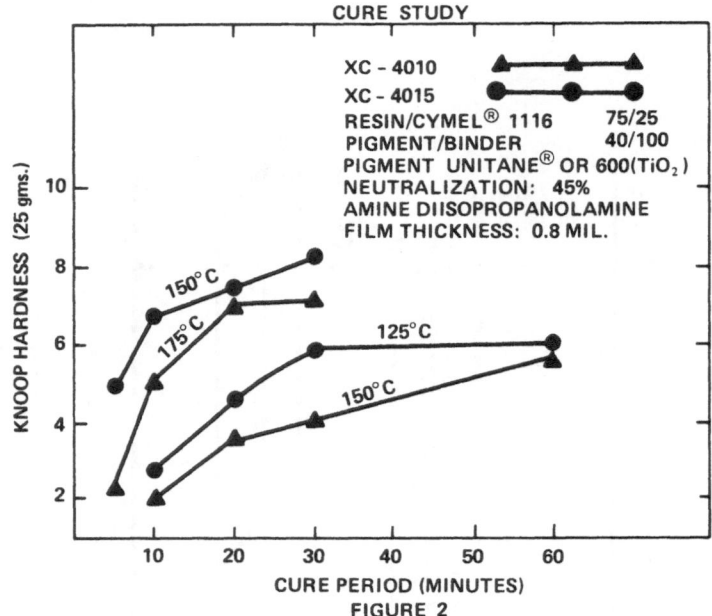

FIGURE 2

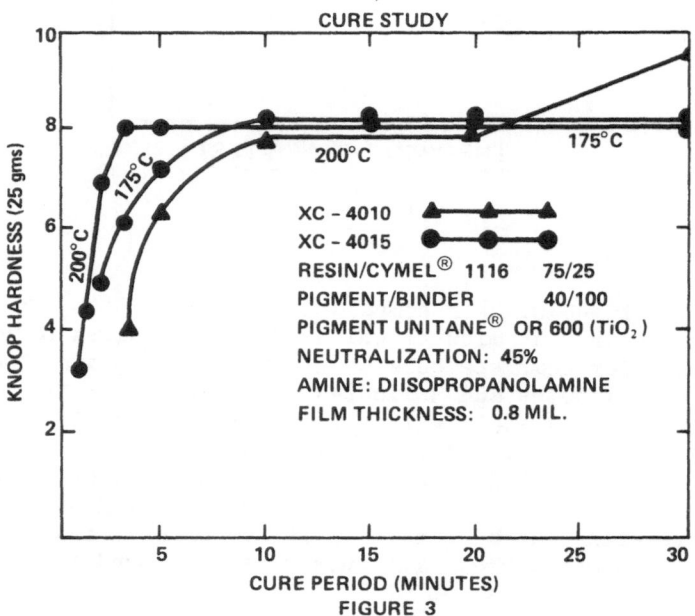

FIGURE 3

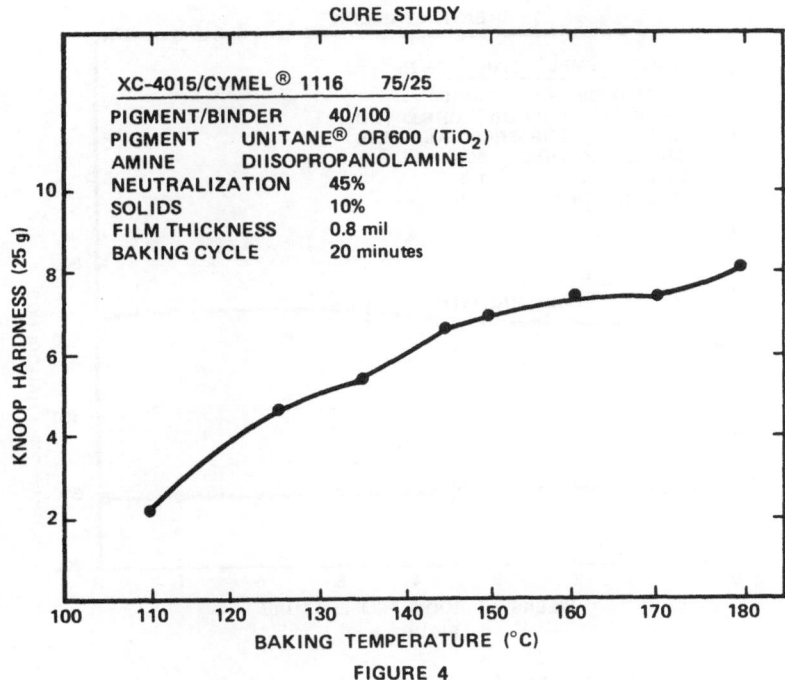

CURE STUDY

FIGURE 4

Cure Rate of XC-4015 as a Function of Temperature

Figure 4 shows the cure study of XC-4015 at various temperatures from 110°C. for a period of 20 minutes.  It shows that the cross-linking reaction is initiated at 110°C but complete cross-linking takes place at 150°C.

Bath Stability

The bath stability of the paint based on XC-4015 resin, studied over a period of 8 weeks, was excellent.  As shown in Figure 5 the change in pH was 0.1 and the increase in conductivity was less than 10% of the original value.  Figure 6 shows the change in gloss of the electrocoated films on steel and aluminum substrates with the aging of the bath. Even after 7 weeks the 60° gloss readings were above 80%. There was some decrease in the 20° gloss.  However, after 8 weeks the addition of 0.5-1% of n-butanol on the total bath liquid almost restored the 20° gloss of the films.  Al-

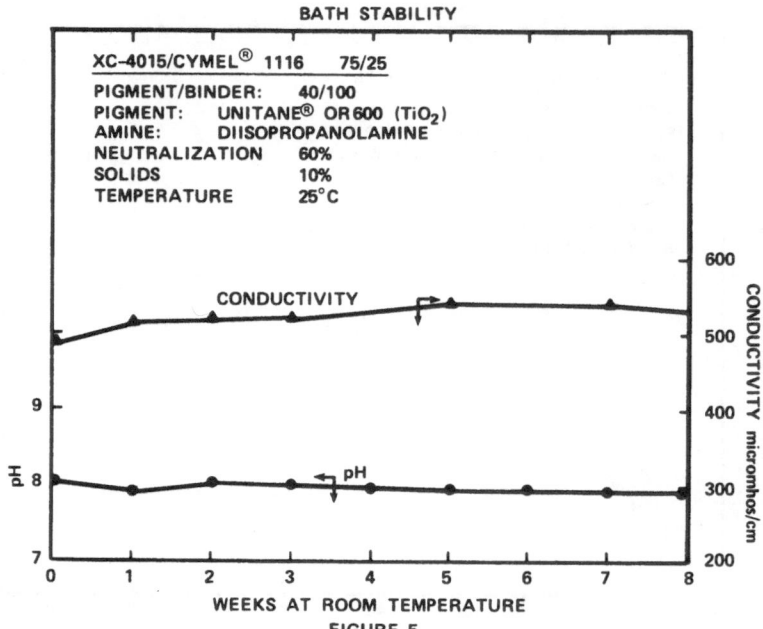

FIGURE 5

so during the aging period there was almost no pigment settling at the bottom of the tank.


## Throwing Power

Figure 7 shows the throwing power of the XC-4015 system.   The throwing power was average.  With an increase in
the level of neutralization a slight drop was observed.   On
aging of the bath there was no significant increase in the
throwing power.


## Corrosion Resistance

The corrosion resistance properties of the XC-4015 re-
sin/CYMEL 1116 amino cross-linking agent were studied on
various substrates.   The electrocoated panels cured at 125°C
for 30 minutes were placed in a salt fog cabinet conforming
to the ASTM specification number B117-64.   The rust creepage
and blistering results obtained after 240 hours are shown in
Table I.

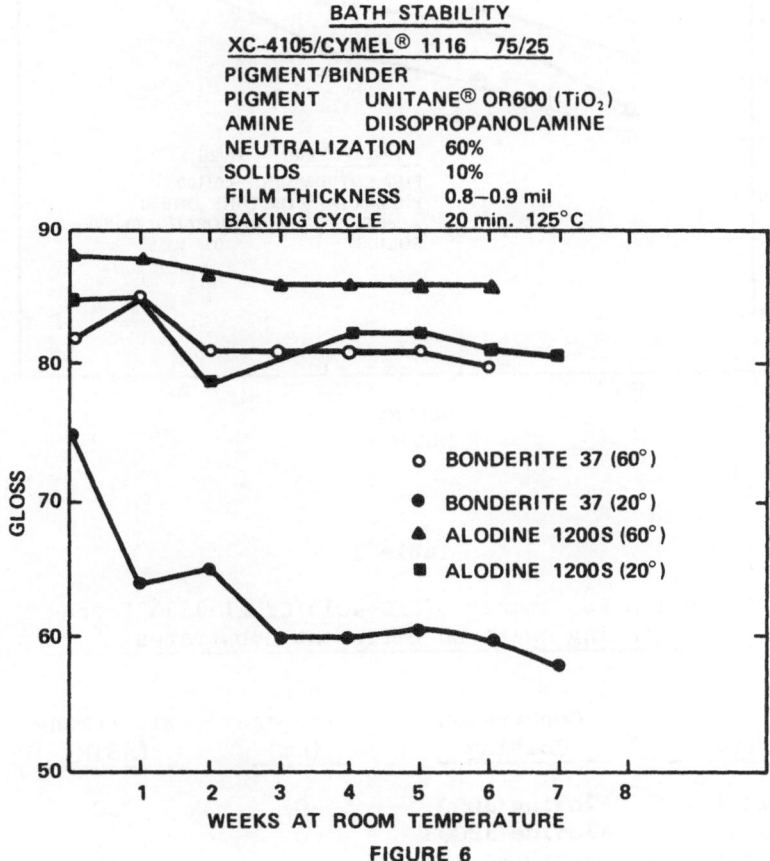

BATH STABILITY

**XC-4105/CYMEL® 1116     75/25**

PIGMENT/BINDER
PIGMENT          UNITANE® OR600 ($TiO_2$)
AMINE            DIISOPROPANOLAMINE
NEUTRALIZATION   60%
SOLIDS           10%
FILM THICKNESS   0.8–0.9 mil
BAKING CYCLE     20 min. 125°C

○   BONDERITE 37 (60°)
●   BONDERITE 37 (20°)
▲   ALODINE 1200S (60°)
■   ALODINE 1200S (20°)

GLOSS

WEEKS AT ROOM TEMPERATURE

FIGURE 6

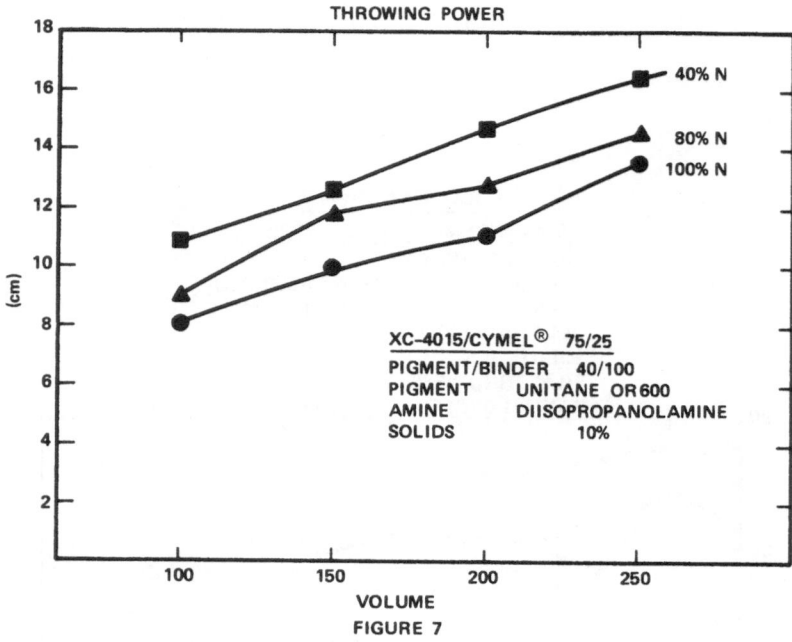

FIGURE 7

Table I

Corrosion Resistance of XC-4015/CYMEL 1116 Cross-
Linking Agent on Different Substrates

| Metal | Conversion Coating | Rust Creepage (mm) | Blistering (ASTM) |
|---|---|---|---|
| Aluminum | Alodine 1000 | 0 | 10 |
| Aluminum | Alodine 1200S | 0 | 10 |
| Steel | Methabond 14* | 1 | 10 |
| Steel | Bonderite 40 | 1 | 10 |
| Steel | Bonderite 823 | 1 | 10 |
| Steel | Bonderite 801 | 5 | 10 |
| Steel | Bonderite 37 | 1 | 10 |
| Steel | Bonderite EP-2 | 0.5 | 10 |

* Lubrizol Corporation, Cleveland, Ohio 44117

## DISCUSSION

The electrocoating and film properties of a new saturated anionic acrylic resin XC-4015 in combination with CYMEL 1116 amino cross-linking agent were studied in detail. The study shows that the electrocoated films cure faster and are non-yellowing. The stability of the bath based on the XC-4015 resin is excellent. The films have good corrosion resistance on suitably pretreated substrate.

There are a number of other ways to achieve a faster or low temperature film cure but with some built-in disadvantages. For example, with the use of an unsaturated resin vehicle, the cured electrocoated films are yellow and have poor alkali resistance. This is due to the fact that the cross-linking reaction takes place through oxidation of the double bonds and the free, untied acid groups of the resin are responsible for the poor alkali resistance of the film.

Saturated anionic resin in combination with a commercially available butylated melamine resin when used in electrocoating results in poor bath stability. The free methylol groups of the butylated melamine resin have the tendency to self-condense under alkaline bath conditions. This condensation reaction accelerates the coagulation of the bath. Our studies[3] indicate that in all cases the bath shows signs of instability in less than 3 days after the preparation of the paint. The addition of a small quantity of a strong acid catalyst to achieve a faster cure is not practical because it results in increased conductivity and poor stability of the bath, and the cured films have poor corrosion resistance.

With the synthesis of the resin XC-4015 and its utilization as a vehicle in electrocoating in combination with CYMEL 1116 cross-linking agent, a faster and a lower temperature cure is achieved without affecting bath stability, deposition characteristics, and film properties. CYMEL 1116 cross-linking agent is very stable under alkaline conditions.

This new resin system shows the best corrosion resistance when deposited on a zinc phosphated steel substrate. Iron phosphated steel substrates or cold rolled steel give poor corrosion resistance. The corrosion resistance decreases with the increase in the amine level of the bath.

This has been explained[2] by a more uncoiled structure of the deposited film with increasing levels of amine in the bath.

In conclusion, with the development of the new XC-4015 resin, fresh interest will be generated in developing faster curing, single white coat electrocoating systems.

## ACKNOWLEDGEMENT

I wish to thank Mr. Werner J. Blank for his helpful suggestions and assistance in preparing this manuscript.

## REFERENCES

1.  J. N. Koral, W. J. Blank, and J. P. Falzone, J. Paint Technol. 40 (519), 156 (1968).

2.  W. J. Blank, J. N. Koral, and J. C. Petropoulos, ibid. 42 (550), 609 (1970).

3.  (G. G. Parekh, February 1971, unpublished data.)

ELECTRODEPOSITION SYSTEMS:  COMPOSITION, CHARACTERISTICS,
PERFORMANCE AND MAINTENANCE

Peter J. Palackdharry

Sherwin-Williams Research Center

10909 Cottage Grove, Chicago, Illinois   60628

ABSTRACT

Electrodeposition systems formulated for specific applications such as single-coat whites over aluminum, appliance
primers, high reflectance whites for steel, high speed
coatings, low bake coatings, automotive primers, cathodic
deposition and a host of other uses vary significantly in
characteristics and performance.  These differences result
directly from the types and amounts of the selected materials
incorporated in a particular formulation.  Once an ED system
has been prepared and its performance satisfactorily checked,
then special precautions and techniques must be applied in
order to maintain the desired properties of the paint in the
coating tank.  This paper is intended to give a general des-
cription of the role and special characteristics of some of
the materials normally used in electrocoating formulations.
Bath stability and throwing power are defined, and the
chemical and physical tests performed to evaluate the durability and appearance of electrodeposited films are specified.  Electrochemical coagulation, electrophoresis,
electrolysis and electroosmosis are also defined.  Some comments are offered on the types of equipment and tank design
needed for commercialization of the process.  Some operational procedures for production are outlined.  Hints are offered
on some of the possible sources of ionic contamination and
the role of ultrafiltration is briefly described.  Finally,
some comments are included on cathodic systems.

## INTRODUCTION

Since the Ford Motor Company installed the first
electrodeposition (ED) system in Europe some eight years
ago, this new and revolutionary painting technique[1] has been
subjected to many intensive and thorough investigations by
both large and small paint manufacturers throughout the
world.  In fact, from the very inception of this process,
its vast potential has been optimistically judged and predicted and subsequent explorations[2] have resulted in a
wealth of information (unfortunately, much of it 'confiden-
tial') and diverse applications.  For example, ED systems
are now being formulated for specific applications such as
single-coat whites over aluminum, appliance primers, whites
for steel, high speed coatings for strappings, coils, etc.,
low-bake coatings, automotive primers and a host of other
uses.  It must be pointed out, however, that the road to
formulating a successful ED system is not in the least bit a
very smooth one, as many of you have probably already
realized.  Many problems, some immediate and others impend-
ing, will have to be solved.  Foremost among these rank the
acceptable application of the coating on different substrates[3]
and the eventual maintenance of a stable paint in the tank[4].

## COMPOSITION AND CHARACTERISTIC

An ED system is generally composed of about 5-15%
solids.  About 95% of the total volatile material is water,
the rest being cosolvents and other volatile additives.  The
solids are composed mainly of the resin, pigment and cross-
linking agent(s).  Many different types of resins are now
available for use in electrocoating systems.  Among these
are epoxy esters, acrylics, polyesters, styrene-maleic anhydride esters, styrene-allyl alcohol esters, maleinized oils,
styrene-butadiene, etc.  The composition of each resin is in
many instances tailored to meet specific needs.  Also, each
resin type is generally best suited for a particular application.  For instance, single-coat whites over aluminum formulations normally contain an acrylic resin while for appliance primers epoxy esters are widely used.  Styrene-allyl
alcohol and epoxy esters are being used for automotive
primers and polyesters and acrylics have been designed for
use in relatively low-bake systems.  It must be remembered,
however, that in many instances a skilled formulator will
find that some appropriate combination of two or three

different types of resins will give the desired performance required for a particular application. Work in our laboratory has indicated that the length and configuration of side chains attached to a polymer back-bone will effect the eventual performance of the vehicle.

Both inorganic and organic pigments are being used in ED systems. Titanium dioxide, of course, finds the most extensive use. In addition, clays and certain other silicates, iron oxide, calcium carbonate, certain chromates and other metal oxides all form part of the available inorganic pigment list. Carbon black, as everyone knows, is also very widely used. Many shading bases also find their way into particular formulations. Experience has shown that special precautions must be adopted to prevent gassing and other manifested detrimental effects when organic pigments are used. But perhaps the greatest difficulty arises in the process to obtain good pigment dispersion in the system in order to eliminate 'kick-out' and gravitational settling. Such a process involves adequate 'wetting' of the pigment by the vehicle. This is normally accomplished by grinding the pigment into the vehicle or by the preparation of pigment 'slurries'. Generally, specific surfactants and other dispersing agents are required for preparing such 'slurries'.

Various different types of cross-linking agents are utilized to achieve maximum film durability and to meet other specific requirements. Alkoxyl melamines and alkoxyl benzoguanamines, phenolic resins, epoxides, styrene-maleic anhydride resins, etc., all have potential uses in this area. Usually the film is cured at specific baking temperatures through transetherification and/or transesterification reactions[5]. The amount and type of cross-linking agent incorporated in a particular formulation is sometimes dependent upon its effect on good pigment dispersion in addition to the desired characteristic of flexibility and durability of the deposited coating as determined by standard tests. Systems have been prepared containing as low as 5% and as high as 50% cross-linker based on the respective total nonvolatile vehicle contents.

Alkanol amines, aliphatic and aromatic amines, and alkali hydroxides are used as solubilizers for practically all anodic ED systems. In many instances, a fixed amount of carboxylic acid groups contained in the polymer are neutralized in order to effect water compatability between

the organic and aqueous phases and also to produce ionized
carboxyl groups that cause migration of the polymeric species
to the anode[6].  Some formulations contain amine above the
amount required to neutralize theoretically all the carboxyl
groups in order to maintain the system on the basic side.
In such cases, where the milliequivalents of solubilizer
per gram or 100 grams of solid is a characteristic parameter
of the formulation, careful checks must be kept on the amine
content as the system is being depleted and reconstituted.
Appropriate steps must be taken to prevent the 'build-up' of
excessive amine in the system either by adding partially
neutralized 'make-up' materials during the reconstitution
process or by some other method.  The use of alkali hydroxide
could present similar problems that would require similar
solutions.

    Cosolvents contained in ED systems play a major role
towards improving film characteristics.  Alcohols, ether-
alcohols, ketone, aromatic and aliphatic hydrocarbons, have
all been used in this respect.  Some experts have claimed
that the selection of a cosolvent should be based on certain
range of values for the solubility of water in a particular
solvent and also the solubility of that solvent in water[7].
However, in addition, perhaps some emphasis should also be
given to the solubility parameter value of the solvent as
compared to that of the resin.  Some preliminary work in our
laboratory has indicated that an effective cosolvent for an
ED system could be a solvent having a solubility parameter
value slightly lower than that of the resin incorporated in
the system.


                    CHARACTERISTICS AND PERFORMANCE

    Once an ED system has been prepared, characteristics
such as pH, specific resistance, throwing power, solids content, milliequivalents solubilizer per gram of solids,
cosolvent level, etc. are immediately established.  In
fact, it is extremely important to establish analytically the
total composition of both the initial and depleted ED paint
in order to facilitate effective reconstitution once the
system has been depleted.  Thus, the migration rates of the
solid components are then known.  One method used for
evaluating the performance of any ED system is the following:  Films are deposited on selected substrates at some
constant time on a voltage ladder beginning at some low

voltage to a voltage at which visual gassing of the system is observed. The coulombic consumption for each deposit is recorded at specific time intervals over the total deposition time span. (Note that certain substrates containing conversion coatings are preheated to remove some water of hydration before deposits are made.) The films are then rinsed and oven-cured in preparation for testing and evaluation.

Some physical tests performed on the cured films are the following: film thickness, gloss, pencil hardness, cross-hatch adhesion, mar resistance, direct impact resistance, reverse impact resistance and conical mandrel impact tests. In addition, the cured films are subjected to salt spray exposure, detergent solution exposure, weatherometer exposure and exterior exposure, all in an effort to evaluate the performance of the coating.

DEPOSITION MECHANISM: ELECTROCHEMICAL COAGULATION

Much detailed work has already been published concerning mechanisms of film formation at the anode during the electrocoating process[2]. Polymers modified with carboxyl functionality and neutralized with bases form water dispersible, negatively charged species which migrate and deposit onto the anodic substrate. Film formation occurs by conversion of the negatively charged carobxyl groups to the acid form[8], by salt formation with metallic ions at the anode or by complexing of more than one functional group with a polyvalent substrate ion. The extent to which any one of these reactions occur depends upon the type and configuration of the polymer contained in the system as well as the chemical and physical nature of the substrate to be coated. Sometimes, aging of a system can result in a higher amount of salt formation than initially observed.

Recently, both in Europe and the U.S., greater effort has been made to design individual polymer systems for different types of substrates in order to achieve the desired and improved performance characteristics of the electrodeposited coating. Metallic ions, produced as a result of anodic dissolution, have been known to reduce the desired properties of coatings[3] and to impart color to light colored systems, reducing both the whiteness and reflectance properties.

DEPOSITION MECHANISM:   ELECTROPHORESIS, ELECTROLYSIS,
ELECTROOSMOSIS AND THROWING POWER

The migration or mobility of particles contained in an
ED system onto the substrate to be coated is defined in
terms of electrophoresis.  Particle charge or zeta potential,
the electrode potential or voltage and the dielectric constant of the medium are the three factors directly related
to mobility while the viscosity of the medium has an inverse
relationship.  It should be noted that relative increases
in zeta potential and the applied voltage will normally
result in an increase in the amount of paint deposited
during a given time period.  The voltage can be controlled
accurately enough so that the resultant film thickness could
be predetermined.  The dielectric constant is determined
primarily by the aqueous medium.  Since viscosity bears an
inverse relationship to mobility, electrocoating baths are
formulated at low solids (5-15% NVM).

Electrolysis occurs as a result of the coating process[2,4].
The surface of the substrate to be coated undergoes oxidation
and ions generated are occluded in the film.  However, aluminum noramlly undergoes anodizing.  Hydrogen ions and oxygen
gas are also generated at the anode.  The main reaction at
the cathode results in the generation of hydrogen gas.
Electroosmosis is a mechanism resulting in the deposited
films being squeezed practically dry of water.  Deposited
films contain as much as 95% solids.

Throwing power is defined as the ability of an ED system
to coat recessed areas of the anode[9].  In general, the lower
the number of coulombs required to form an insulated film
over a relatively wide voltage range and the greater the
bath conductivity, the higher is the throw.  During the
initial stages of deposition the outer edges of a complex
profile are first coated.  The current flow occurs through
the paths of least resistance.  Thus, as the film builds on
the outer edges and profiles and the coated parts become
relatively insulated, the paths of lower resistance are then
shifted further into the recessed areas of the electrode.
Eventually, within a relatively very short time, a film
deposit of uniform thickness is obtained on all exposed
areas.  It should be pointed out that the development of ED
systems with higher throwing power was the major factor
responsible for the commercialization of this process in

the automotive industry. The prior need for complicated
auxiliary electrodes has practically vanished.


## EQUIPMENT AND OPERATIONS

The following descriptions of equipment and operational
procedures are mostly a part of an aluminum extrusion line
presently in operation. Aluminum extrusions are pinned
onto racks in readiness for the pretreatment process whereby
the surfaces to be coated are thoroughly cleaned. The parts
are then thoroughly rinsed with deionized water immediately
before being immersed in the paint tank so that contamination of the paint is practically avoided. Usually, water
of resistance greater than 100,000 ohm-cm is required for
this rinsing process. The parts are then electrocoated.
Gaseous products such as hydrogen are collected in a weir
compartment and effectively dissipated without becoming
entrapped in the paint film.

The DC power supply unit is an integral part of the
electrocoating line. Presently, most new electrocoating
installations are completely automated. In addition, if
increased amperage capacity is needed at some future date,
it can easily be achieved by incorporating additional
modules to the power supply unit.

Circulation of the paint in the tank is achieved either
by the use of vertical pumps or draft tube agitators. Since
temperature changes will affect the electrical conductivity
of the paint, heaters and heat exchanger units are widely
utilized for temperature control. Premix tanks, containing
both paint from the coating tank and solubilizer deficient
'make-up' materials are agitated by high speed, high shear
mixers. The premix tanks are in line with the coating tank
so that replenishing of the depleted bath is readily
accomplished. After the parts are coated, there still
remains a thin layer of 'undeposited' paint or 'drag-out'
that must be removed prior to baking. This is accomplished
by the use of immersion rinses. It should be pointed out
that most lines are equipped with facilities to coagulate
and extract the paint from the immersion rinse tanks,
thereby cleaning the effluent water. After rinsing of the
coated parts is accomplished, the ware is ready for baking.
Practically all present electrocoating finishes require a
bake for curing.

## SOURCE OF PROBLEMS

Many difficulties could be experienced with ED systems operating in the field as a result of changing conditions within the bath. For example, degradation of the vehicle could occur as a result of hydrolysis. Crosslinking reactions could result in a build-up of unacceptable amounts of extremely high molecular weight materials. Excessive temperature fluctuations could result in an unstable bath. The solubilizer and cosolvent levels must be maintained within a tolerable limit. Major difficulties could easily develop as a result of ionic contamination. Calcium, iron, chromium, sulfate, chloride and a host of other ionic species are detrimental to the stability and efficiency of ED baths. Calcium ions in concentrations as low as 10-20 ppm on the total bath and sulfate ions in about 50 ppm have produced unacceptable coatings from a bath. Thus, it has now been generally recognized that ionic contamination of ED systems must be maintained to as low a level as possible (or possibly be eliminated) in order for the process to work with any reasonable degree of effectiveness.

Therefore, careful consideration should be given to the installation of ultrafiltration units as an integral part of any ED operation, especially where ionic contamination is expected to be a major problem. In addition, quality control checks should be constantly made on all raw materials incorporated in any specific formulation. Efficient ion-exchange columns must be used when deionized water is added to the bath. Finally, utmost care must be exercised at all times to prevent any of the pretreatment cleaning solutions from entering the bath as the substrates are being prepared for electrocoating.

## ULTRAFILTRATION

Recently, many articles have been published in current journals detailing the basic principles and merits of ultrafiltration technology[10]. This technology was found to be of invaluable use in the electrocoating area. Ultrafiltration techniques are being employed to extract water soluble contaminants from electrocoating baths. In addition, relatively low molecular weight organic species produced as a result of degradation or cross-linking reactions could also be extracted. Excess solubilizer and/or cosolvents are

readily removed from an ED bath by this process. It should
also be noted that it is now possible to collect enough fil-
trate by this process for use in rinsing the 'drag-out'
materials off the freshly coated parts and return this paint
to the coating tank. Thus, a closed loop post-rinse station
is established which assures nearly 100% utilization of the
paint solids[11] and precludes the need for flocculating equipment.

## CATHODIC DEPOSITION

The development and application of cathodic systems
for coating metallic substrates represent a natural exten-
sion of the intense work done on conventional anodic
systems. This feat has been accomplished by the proper
tailoring of polymers to deposit at the cathode.

With the use of cathodic systems, substrate ion contamination of coatings as a result of anodic dissolution
will not occur. Thus, the performance of coatings should
be improved. In addition, polymer types possessing except-
ional properties could be developed and employed that cannot
be normally used in anodic systems. One of the major con-
cerns in the development of cathodic systems is the resulting acidic nature of the baths whenever acids are used to
produce the positively charged species. These paints could
be highly acidic and problems in both tank and conveyor
design would most probably be encountered. However, such
problems are not specific for all cationic systems. Some
polymers are being designed to overcome this disadvantage.
Finally, the question would have to be settled as to what
materials could be effectively used as the anode so that
the paint would not be contaminated with anodic dissolution
by-products.

## NON-POLLUTION ASPECTS OF ELECTROCOATING

It has already been mentioned that an electrodeposition
system noramlly contains a maximum of about 3-4% cosolvent(s)
in the total bath. Accumulated evidence indicates that very
little (if any) of the cosolvent(s) is lost by evaporation.
Thus, there is no known violation of Rule 66. In addition,
with the importance of ecology in mind, the rinsed-off 'drag-
out' materials from the electrodeposited coating are

coagulated and the solid portion extracted, thereby cleaning
the effluent water. Finally, with the availability of
ultrafiltration, the rinsed-off 'drag-out' materials can be
recycled into the system, thus assuring nearly 100% utilization of the paint and thus precludes the need for flocculating equipment. All these factors contribute toward
labeling electrocoating as a non-polluting coating process.


ACKNOWLEDGEMENT

The author gratefully wishes to acknowledge the help
afforded to him by all members of the Electrocoating Team
at the Sherwin-Williams Company, Chicago, Illinois.


REFERENCES

1.  Bogart, H. N., Burnside, G. L. and Brewer, G. E. F.,
    "The Concept of Development of the Ford Electrocoating
    System", Society of Automotive Engineers, Paper 988A.

2.  LeBras, L. R., "Electrodeposition - Theory and Mechanism",
    J. of Paint Technology 38, No. 493, Feb., 1966.

3.  May, C. A. and Smith, G., "Dissolution of the Anode
    During the Electrodeposition of Surface Coatings", J. of
    Paint Technology 40, No. 526, Nov., 1968.

4.  Hays, D. R. and White, C. S., "Electrodeposition of
    Paint: Deposition Parameters", J. of Paint Technology
    41, No. 535, Aug., 1969.

5.  Blank, W. J., Koral, J. N. and Petropoulos, J. C., J. of
    Paint Technology 42, No. 550 (1970).

6.  Rheineck. A. E. and Usmani, A. M., J. of Paint Technology
    41, No. 597 (1969).

7.  U.S. Patent 3,434,592.

8.  Tawn, A. H. R. and Berry, I. R., JOCCA 48, 790 (1965).

9.  Brewer, G. E. P., Strosberg, G. G. and Horsch, M. E.,
    J. of Paint Technology 42, 550 (1970).

10.  Goldsmith, R. L., Ind. Eng. Chem. Fundam. 10, 1 (1971).

11.  LeBras, R. L. and Christenson, R. M., J. of Paint Technology 44, 566 (1972).

# ULTRAFILTRATION OF ELECTROCOATING SYSTEMS

Arnold J. Josefson and Loren R. Munson

DeSoto, Inc.
1700 South Mt. Prospect Road
Des Plaines, Illinois 60018

## INTRODUCTION

Ultrafiltration, a process for separating molecular dispersions on the basis of molecular size, has recently found application in the area of electrocoating. Because of potential paint savings and the possibility of tank and pollution control, it has made an immediate impact on both suppliers and users of electrocoating paint.

Ultrafiltration is a membrane separation process. The separation is effected by means of a semipermeable membrane. These membranes are constructed with very small pores so that small molecules and ions pass through the membrane while particles of larger dimension, such as pigment and resin particles, are retained. A thorough discussion of the theoretical aspects of the process can be found in the literature. [1,2,3,4]. However, very little concerning the practical application of ultrafiltration to the electrocoating process has been written. The aspects of the process that are pertinent to this study are those which relate to the bath parameters; concentration, temperature, composition, and the ultrafiltration operating parameters; pressure and flow rate.

## EXPERIMENTAL

All experiments were performed using cellulosic membranes of approximately 2.2 ft.² total surface area. Circulation was provided by a 1-1/2 hp. centrifugal pump. The flow rate of the mainstrem was determined by measuring the pressure drop across a calibrated orifice. The maximum mainstream flow rate obtainable was approximately 14 gpm at a pressure of 19.5 psig. The maximum pressure obtainable was 30 psig. Permeate flow rates were determined by measuring the elution time for a specified volume. A diagram of the apparatus can be seen in Figure 1.

The majority of the experiments were run using a nonpigmented system based on an acrylic resin. Other parts of the study were performed using acrylic or epoxy ester type vehicles. The analyses of the baths and permeates were performed using the methods reported by Anderson and Tessari[5].

## DISCUSSION

A. Operational Variables Affecting the Flux Rate
   1. Pressure

Figure 2 illustrates the effect of pressure on the flux rate, $J$, the rate that small molecules pass through the membrane. Note that for water, $J$ is directly proportional to $P$. However, for the paint, at 8% NV and a flow rate of 7.5 gpm, there is little increase in $J$ beyond 8 psig. When the flow rate is increased to 9.0 gpm, $J$ is increased by increasing the pressure up to 15 psig. By plotting $J$ vs. $P$ at higher flow rates, it is possible to generate a family of curves. The optimum pressure will be a function of the flow rate. This relation between pressure and flux rate can be explained in terms of the effect of pressure on the "polarization layer".

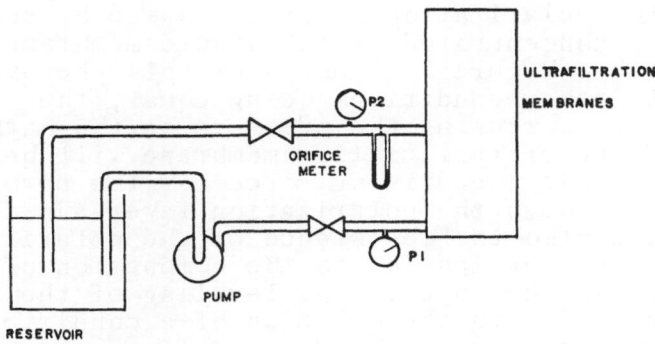

Figure 1 - Schematic of Apparatus

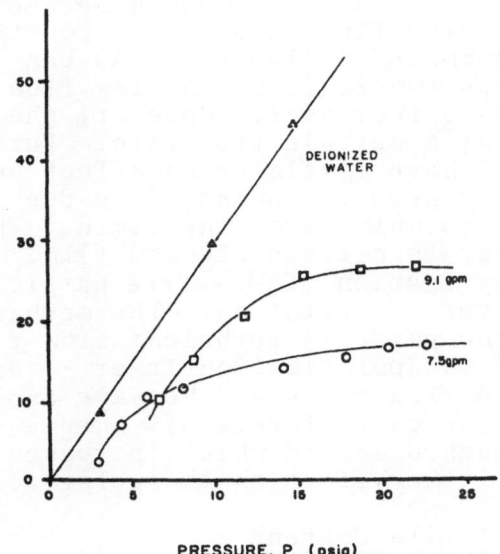

PRESSURE, P (psig)

Figure 2 - Effect of Pressure on Flux Rate for Deionized
Water and a Pigmented Modified Acrylic-Melamine
System

The polarization layer is caused by solids becoming concentrated at the liquid-membrane interface. Figure 3 illustrates this phenomenon. With all other conditions being equal, the effect of increasing the pressure on the bath side of the ultrafiltration membrane will be to both increase the driving force for the permeate to pass through the polarization layer and membrane and also the resistance of the polarization layer to permeation due to the compaction of the material in the layer. The leveling of the J vs. P curve is due to the creation of a condition where an increase in the driving force is almost exactly opposed by an increase of the resistance to permeation of the polarization layer.

### 2. Flow Rate

A plot of the relationship between flow rate, Q, and the flux rate, J, at constant pressure is shown in Figure 4. As can be seen, the curve has two regions. At low flow rate, J increases as Q increases. However, the curve levels off at a certain flow rate. Further increases in Q have little or no effect on J. The increase in J as Q is increased is due to the increase in turbulence of the fluid. The polarization layer is not a coalesced film. Rather, it is an aggregation of discrete particles which can be removed by agitation. The mechanical action of the paint in turbulent flow removes portions of the polarization layer - resulting in increased flux rates at the same applied pressure. The curve levels off when a certain point is reached beyond which increased turbulence has little effect on the polarized layer.

### 3. Nonvolatile Content

The thickness of the polarized layer depends not only on the amount of turbulence but also on the amount of material which is available to reform the layer as it is washed away. Therefore, we would expect that as the amount of resin in the bath is reduced, the thickness of the

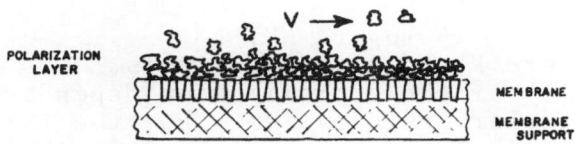

Figure 3 - Representation of Polarization Layer

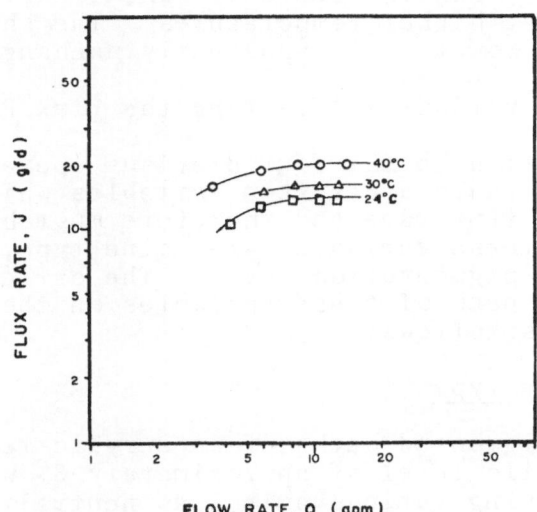

FLOW RATE, Q (gpm)

Figure 4 - Effect of Flow Rate on Flux Rate at Three
Temperatures for a Modified Acrylic System

polarization layer would also be reduced with a corresponding increase in the flux rate. Figure 5 illustrates this relationship. For the range of nonvolatiles shown, the relationship can be considered approximately linear when plotted on a semi-log scale. Mathematically, it can be expressed as:

$$J = K - A \log (\% \text{ NV})$$

where K and A are constants.
The validity of the approximation depends on each particular resin system. For the three systems shown in Figure 4, it is felt that the approximation is very good in the range shown.

### 4. Temperature

Increasing the temperature of the paint decreases the viscosity and in turn increases the turbulence of the paint in the membranes. The effect of varying the temperature or the flux rate can be seen in Figure 4. The flux rate is increased at higher temperatures. The shape of the curve, however, is apparently unchanged.

### B. Systems Variables Affecting the Flux Rate

In addition to the four previous operational variables, there are system variables which can affect the flux rate and therefore must be considered. These variables are amine type, amine level, and pigmentation level. The effects of changes in each of these variables on the flux rate are as follows:

### 1. Amine Type

Nonpigmented baths of an acrylic resin at a nonvolatile level of approximately 8% were prepared using various amines as neutralizing agents. The amines used were diisopropanol amine, triethyl amine, triethanol amine, and dimethyl ethanol amine. The data is presented in Table I. As can be seen, the choice of amine can influence the flux rate.

## 2. Amine Level

The amount of amine present in the bath will also affect the flux rate. Within the limits of the solubility of the resin system, the flux rate will increase as the degree of neutralization is decreased. The flux rate of a system at varying degrees of neutralization are shown in Table II.

Viscosity data obtained from baths of this system at the various neutralization levels illustrates the effect of the percent neutralization on the viscosity of the system and thus on the flux rate. The data was obtained using a Brookfield RTV Viscometer with a #1 spindle and 50 rpm and is given in Table II.

## 3. Pigment Level

In order to determine the effect of pigment loading on the flux rate of a system, five systems were prepared with pigment to binder, (P/B), ratios of 0, .1/1, .2/1, .4/1, and .8/1. The effect of these changes on the flux rate, percent nonvoltaile passage, and bath viscosity are shown in Table III.

The flux rate increases as the P/B is increased. This can be partially explained by the response of the viscosity to the change in P/B. The reduction in viscosity would result in greater turbulence in the flow channel and thus affect the formation of the polarization layer.

## C. Permeate Composition

The composition of the permeate from an ultrafiltration unit is dependent on the composition of the bath that is being ultrafiltered. In general, the permeate will consist of water, solvents, amines, ionic species,and low molecular weight resinous materials. A more comprehensive discussion of each component follows.

TABLE I: THE EFFECT OF AMINE TYPE ON FLUX RATE

| AMINE | NEUTRALIZATION LEVEL (%) | TEMPERATURE (°C) | PRESSURE (PSIG) | FLOW RATE (GPM) | FLUX RATE (GFD) |
|---|---|---|---|---|---|
| Triethylamine | 100 | 30 | 19 | 13.42 | 14.9 |
| N,N-Dimethyl Ethanolamine | 100 | 30 | 19 | 13.44 | 15.1 |
| Diisopropanol-amine | 100 | 30 | 19 | 13.62 | 16.1 |
| Triethanol-amine | 100 | 30 | 19 | 13.55 | 19.1 |

TABLE II: EFFECT OF DEGREE OF NEUTRALIZATION ON FLUX RATE

| SYSTEM | AMINE | % NEUTRALIZATION | TEMPERATURE | PRESSURE | VISCOSITY | FLUX RATE |
|---|---|---|---|---|---|---|
| Modified Acrylic | N,N-Dimethyl Ethanol Amine | 100 | 30°C | 18 psig | 131.4 cp | 1.53 |
| " | " | 80 | 30°C | 18 psig | 104.6 cp | 2.16 |
| " | " | 60 | 30°C | 18 psig | 12.8 cp | 10.29 |

TABLE III: EFFECT OF PIGMENT LOADING

| P/B | %NV | TEMPERATURE | PRESSURE | FLOW RATE | VISCOSITY | FLUX RATE | %NV PASSAGE |
|---|---|---|---|---|---|---|---|
| 0 | 6.0 | 30°C | 20 psig | 11.7gpm | 14.5 cp | 16.6 | 3.3 |
| .1/1 | 6.2 | 30°C | 20 psig | 11.2gpm | 14.2 cp | 17.8 | 3.2 |
| .2/1 | 6.0 | 30°C | 20 psig | 11.2gpm | 14.0 cp | 19.8 | 3.0 |
| .4/1 | 5.9 | 30°C | 20 psig | 11.6gpm | 13.9 cp | 23.4 | 2.5 |
| .8/1 | 6.1 | 30°C | 20 psig | 11.3gpm | 12.6 cp | 24.5 | 2.3 |

*Measured with Brookfield RTV Viscometer: #1 spindle, 100 rpm at temperature indicated.

TABLE IV - EFFECT OF PERCENT NEUTRALIZATION ON PERCENT PASSAGE OF AMINE

| RESIN | % NEUTRALIZATION | AMINE | % PASSAGE |
|---|---|---|---|
| Modified Acrylic | 100 | N,N-Dimethyl ethanol amine | 13.9 |
| | 80 | | 11.4 |
| | 60 | | 10.3 |

## 1. Amines

The amount of amine present in the permeate will depend on the concentration of the amine in the bath and the degree of association of the amine with the resin. The degree of association of the amine with the resin should correspond to the bascisity, $pK_b$, of the amine. The relationship between the $pK_b$ and percent passage for various amines is shown in Figure 6. As can be seen, there is a strong tendency for weaker (higher $pK_b$) amines to pass through the membranes.

The effect of reducing the percent neutralization of the bath on the percent passage of the amine is shown in Table IV. The percent passage is lower for baths prepared at lower neutralization levels.

## 2. Solvents

Since solvents are generally low molecular weight species, it was expected that they would permeate through the membrane. This is indeed the case for the water soluble solvents. Examination of permeate samples by gas chromatography indicates that water soluble solvents such as the glycol ethers and low molecular weight alcohols are found in the permeate in approximately the same concentration as is found in the bath. Non water soluble solvents such as n-dibutyl-ether, however, were not found in the permeate.

## 3. Ionic Contaminants

It has been reported by Philips [6] that the permeate from commercial installations can contain various ionic contaminants such as chromate, chloride, and phosphate ions. These contaminants are a result of treatment of the ware such as phosphatizing or acid pickling prior to entry into the electrocoating tank. Thus, ultrafiltration provides a convenient means of removing

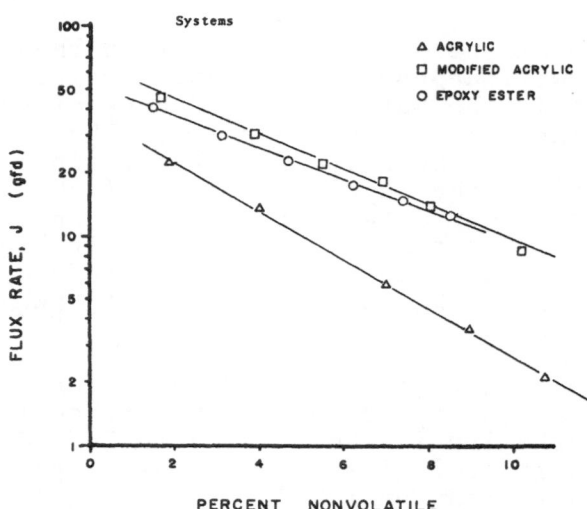

Figure 5 - Effect of Non-Volatile Level on Flux Rate for
Three Systems

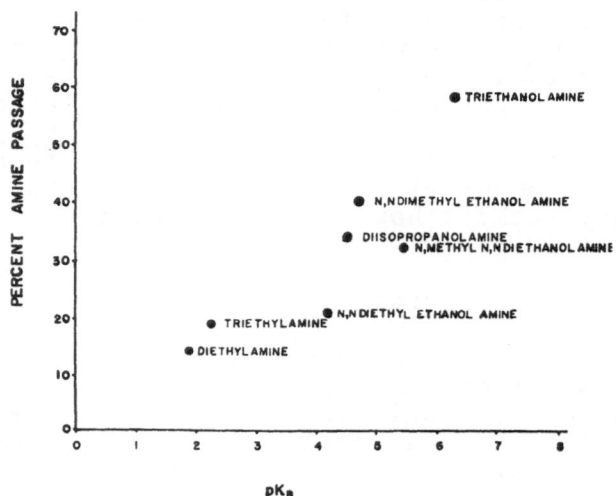

Figure 6 - Effect of $pK_b$ on Percent Amine Passage for
Various Amines

undesirable contaminants from the electrocoating bath

### 4. Low Molecular Weight Resin Fraction

Part of the permeate is composed of non-volatile materials. The amount of nonvolatile portion of the permeate varies widely with the various resins. For the systems used in the ultrafiltration study, the range is from .2% to 4.5% of the bath nonvolatiles. The amount that passes through the membrane is dependent on the solids level of the bath with a greater percentage passing through at the lower nonvolatile levels.

In systems that contain melamine, a greater amount of nonvolatile material is found in the permeate than for the same system without melamine. The following data from an acrylic/melamine system confirms this.

| % Melamine | % Passage |
|------------|-----------|
| 0 | .88 |
| 10 | 1.98 |
| 20 | 2.69 |

Also, analysis of permeate from melamine containing systems by gel permeation chromatography and infrared spectroscopy indicate that the main constituent is melamine. A gel permeation chromatography curve for permeate from a typical acrylic melamine system is presented superimposed on the curve for the electrocoating bath in Figure 7. The IR curve for the same permeate sample is presented in Figure 6.

### CONCLUSIONS

The variables governing the ultrafiltration flux rates of electrocoating formulations have been separated into operational variables and system variables. The operational variables such as temperature, pressure, flow rate, and nonvolatile level are more or less out of the control of a paint supplier. However, system

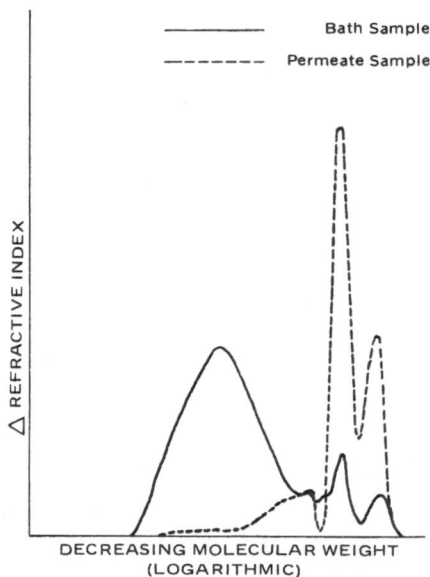

Figure 7 - Gel Permeation Chromatography Curves for Bath
and Permeat from Acrylic - Melamine System

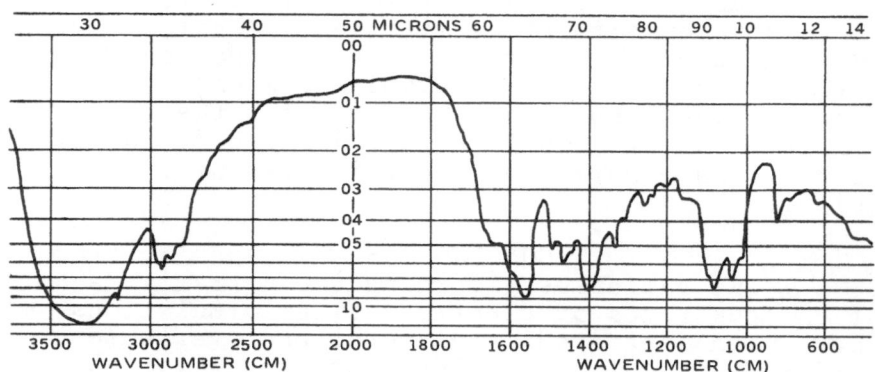

Figure 8 - IR Spectroscopy Curve for Permeate from Acrylic
Melamine System

variables such as amine type, amine level, pigmentation level and resin molecular weight are directly controllable by the paint supplier.

It has been shown that the viscosity of the electrocoating formulation is the controlling factor in determining the flux rate of an electrocoating paint. The effect of the system variables is to change the viscosity of the paint and to thus affect the flux rate. The composition of the ultrafiltration permeate has been determined. The main constituents are water, amine, solvents, low molecular weight polymer fractions, and ionic contaminants.

## REFERENCES

1.    Forbes, F., Product Finishing, 23 (11). 1971, pp. 24-29.

2.    Kesting, R.E., Synthetic Polymeric Membranes, McGraw Hill Book Company, 1971.

3.    Perry, E.S., Advances in Separation and Purifications, John Wiley & Sons, 1968, pp. 143-186, 297-335.

4.    Porter, M.C., and Michaels, A.S., Chem. Tech., Jan. 1971, pp. 56-63.

5.    Anderson, D.G., and Tessari D.J., Journal of Paint Technology, 42, No. 541, (1970), pp. 119.

6.    Philips, G. Electrocoating-An Answer to Pollution Problems? Paper presented to Electrocoat '71 Conference in London.

# QUANTITATIVE PREDICTION OF RECYCLING EFFICIENCY OF ELECTRO-COATING PAINT SOLIDS THROUGH ULTRAFILTRATE RINSE

George E. F. Brewer [*]

Ford Motor Company

24500 Glendale Ave., Detroit, Mich. 48239

## INTRODUCTION

Corrosion of automobile bodies became a severe problem in the 1950's, due to the increased use of salt for the de-icing of roads. The essentially flat outer surfaces of an automobile body were reasonably well protected against the corrosive action of salt and environment through a paint coat. The automotive bodies however corroded from the inside out. A paint application process was needed which would protect both inner and outer surfaces with a uniformly thin paint film.

Dip coating provides sufficient contact between the paint bath and the inner surfaces of merchandise, but the so produced films are uneven in thickness. Films which are too thin lack corrosion protection and films which are too thick cannot be cured completely and tend to lift off. Worse than this, a dip coating bath contains about 50 wt. % of paint solids and 50 wt. % of a liquid carrier, be it water or be it organic solvent, which has to be vaporized during the paint bake operation. During this bake, the outer skin of the merchandise is heated and temperature differences are created between inner surfaces. Thus, the paint solvent evaporates from relatively hot sections and condenses on the cooler surfaces, where it washes away the paint. This phenomenon is called "reflux damage."

[*]Present Address: Coating Consultant, 11065 East Grand River Road, Brighton, Mich. 48116

## ANTIPOLLUTION ASPECTS OF ELECTROCOATING

To avoid reflux damage, that is to obtain the same full corrosion protection on both inner and outer surfaces of merchandise, a paint application process is needed by which paint solids are deposited in virtual absence of any solvent. A paint process which operates in absence of organic solvents is also extremely desirable, because of the absence of emission of air pollutants.

The concept of such a low polluting – high corrosion protecting coating process, called "Electrocoating," originated in the late 1950's at Ford Motor Company[1]. In addition to virtual absence of organic vapor emission during the painting and baking operation, the process also results in much improved corrosion protection and in sizeable cost savings when compared with spray or dip coating[2].


## THE ELECTROCOATING PROCESS

Practically all known film formers, such as epoxy, alkyd, or acrylic resins can be synthesized to contain acidic or basic groups. Thus, if we symbolize the resinous part plus the acidic group by "R COOH," we have a water indispersible, acidic resin. Through reaction with a base (BOH) we receive a water dispersible macro-anion "RCOO⁻" and its counter-cation "B⁺"[3].

If two electrodes are immersed into such an aqueous dispersion, the film forming macro-anion will be deposited on the positive electrode. The deposition will start on the electrically most exposed area. If the deposited film exhibits high electrical resistance, it blocks further flow of electricity, the current will seek the nearest still available path and in about one minute all metal surfaces of such a complex structure as an automotive body will be coated with a paint film of uniform and desirable thickness. The electrocoating process is currently practiced world wide in about 500 major installations for the corrosion protection of merchandise, such as steel trusses, automotive bodies and components, appliances, furniture, toys and even nuts and bolts[4].

## ELECTROCOAT RINSE STAGE

Merchandise with its freshly deposited electrocoat is lifted from the coating tank. The coating bath – having approximately the viscosity of water – drains fast from the workpiece. There are however bath droplets adhering to the electrodeposit, and – depending on the design of the merchandise – puddles of coating bath may fill nooks and corners.

As an example, about 2,000 g of paint solids are required for the electrocoating of an average automobile body. It is estimated that the body carries with it about 8,000 g of electrocoating bath, which, at 10% bath solids, equals about 800 g of paint solids.

These paint solids have to be rinsed away, since – on evaporation of the bath water during the paint bake – each bath droplet will leave a minute, but much visible crut on the painted surface. Also, the evaporation of a bath puddle would leave a thick layer of paint, which may not give the desired corrosion protection.

The design of the electrocoat rinse stage makes use of the well developed technology of metal cleaning and pretreatment. Workpieces are moved over an open tank which contains rinse fluid. A pump moves the desired volume of rinse fluid to nozzles (see Figure 1). Fresh rinse fluid is supplied to the tank, while the used fluid overflows into a waste treatment system.

Let us define a "complete rinse" as the operation under such conditions that only a tolerable amount of impurity remains on the rinsed merchandise. In order to use the minimum amount of rinse fluid, the operation should be carried out so that the rinse fluid remaining on the merchandise should contain practically the same concentration of impurity as the spent rinse fluid which leaves the rinse deck.

In the case of automotive bodies, a reasonably complete rinse is achieved through the impingement of about 300 liter of water per minute for approximately one minute.

The rinse water consumption can be drastically cut when applied in a counterflow, cascading rinse system.

# RINSE DECK

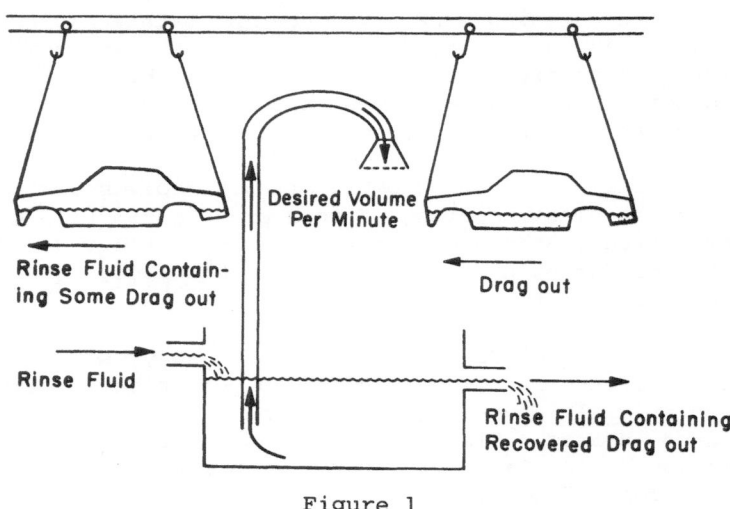

Figure 1

It is not difficult to keep the rinsed off electrocoating paint solids in re-usable form, cut it is not possible to run even small volumes of rinse fluid into the electrocoating tank without overflowing.

## ULTRAFILTRATION OF ELECTROCOATING BATH

Recent advances in technology have provided us with
membranes capable of retaining dispersed colloids and pigments, while passing truly dissolved materials (see Figure
2) and insuring defect free operation for one or more years.
These ultra filters of varying designs, are capable of de-
livering about 1 liter per minute per 2 square meter (1 gpm
per 75 sq. ft.).

### ULTRAFILTRATION

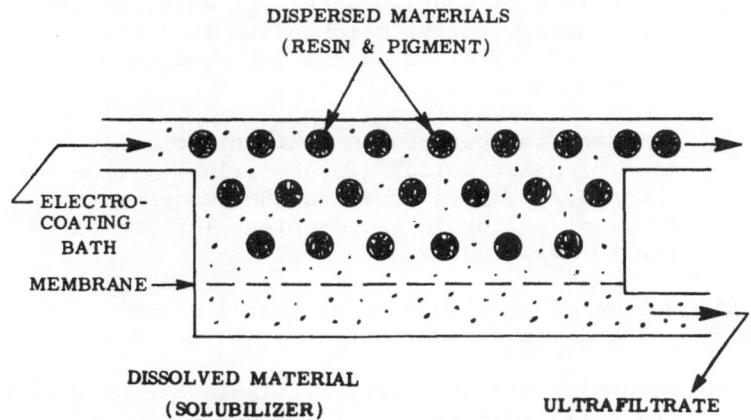

Figure 2

100 g of typical ultrafiltrate from an electrocoating bath contains approximately 0.5 g paint solids, 0.3 g of paint solubilizer, and small amounts of various dissolved substances. In all, the ultrafiltrate is the habitat from which dispersed paint solids are electrodeposited. The ultrafiltrate is, therefore, an excellent fluid for the recycling of carried-out bath droplets, and for their return to the paint bath.

Thus, bath fluid can be ultrafiltered from an electrocoating bath, and, after use as rinse fluid, be returned to the bath.

COMPUTATION OF PAINT RECOVERY THROUGH 1-STAGE RINSE

Figure 1, indicated that a rinse deck is a system of four streams: Rinse fluid and dragout enter, while rinse fluid containing some dragout, and rinse fluid containing recovered dragout leave.

Two reasonable assumptions can be made:

1.  The volume of coating bath ($V_b$) entering the rinse deck with the merchandise is equal to the volume of fluid on the merchandise when leaving the rinse deck.

2.  The deck is operated so that the concentration of paint solids ($C_1$) overflowing from the rinse tank is equal to the concentration of paint solids in the droplets left on the rinsed merchandise.

The volume of coating bath on the merchandise ($V_b$) is determinable, and its solids concentration is known ($C_b$).

The volume of ultrafiltrate available per unit time ($V_{uf}$) and its solid contents ($C_{uf}$) are known. Thus, there is only one unknown: ($C_1$).

Suppose now that an open tank of a certain volume ($V_t$) is filled with ultrafiltrate, and the rinse operation begins. Obviously the first piece receives the best rinse, since ($C_1$) will be lowest. Eventually, however, a steady state will be reached, that is, ($C_1$) will attain a maximum value. From there on the concentration ($C_1$) will not change with time:

$$V_t \frac{dC_1}{dt} = 0$$

From that time on, the paint solids carried into the rinse system will be equal to the quantity leaving the tank, or

$$(V_{uf})(C_{uf}) + (V_b)(C_b) = (V_b)(C_1) + (V_{uf})(C_1)$$

The left side of the equation represents the incoming quantity of paint solids, and the right side the outgoing quantities.

Figure 3 shows typical data. In this particular case, the paint solid recovery will be a minimum of 456 g per minute (57% recovery) when steady state is reached. There will, however, be higher recovery before equilibrium is reached.


COMPUTATION OF PAINT RECOVERY THROUGH MULTI-STAGE RINSE

The reclamation of dragged out coating bath through ultrafiltration is commercially well justified. The investment in ultrafiltration equipment is considerable and ultrafiltrate should therefore be used efficiently. This can be done in counterflow, cascading multi-stage installations.

The rinse deck shown in Figure 3 can be used in series with another stage. For simplicity let us consider a second stage with identical tank size and spray rate, as shown in Figure 4. While 12 liter ultrafiltrate per minute in a single stage rinse brought a 57% recovery, the same 12 lpm give 75% recovery in a 2-stage rinse.

Figure 5 extends the same parameters to a 3-stage rinse with resulting 83% minimum recovery.

## I-STAGE RINSE

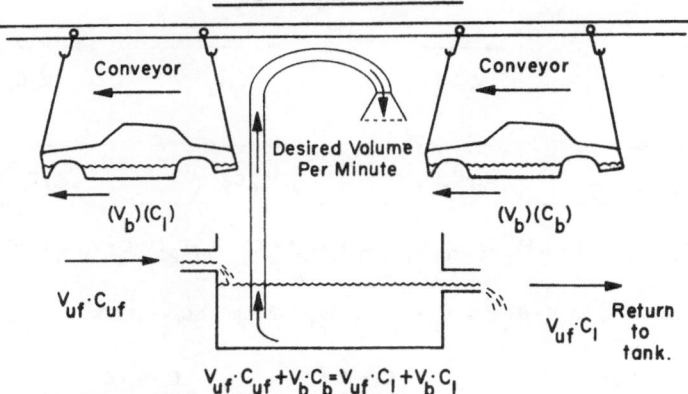

$$V_{uf} \cdot C_{uf} + V_b \cdot C_b = V_{uf} \cdot C_l + V_b \cdot C_l$$

EXAMPLE: $V_b = 8$ lpm ; $C_b = 10\%$ ; $V_{uf} = 12$ lpm ; $C_{uf} = 0.5\%$.

$C_l = 4.3\%$

57 % Recovery

Figure 3

## 2-STAGE RINSE

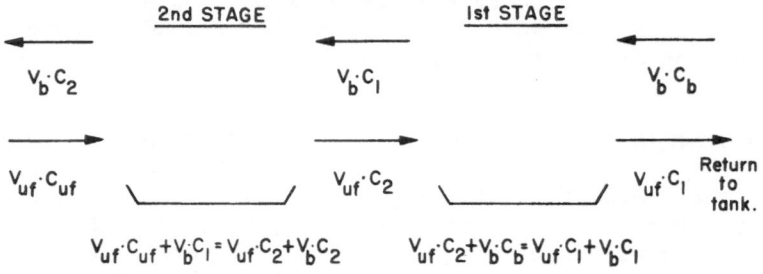

$$V_{uf} \cdot C_{uf} + V_b \cdot C_1 = V_{uf} \cdot C_2 + V_b \cdot C_2 \qquad V_{uf} \cdot C_2 + V_b \cdot C_b = V_{uf} \cdot C_1 + V_b \cdot C_1$$

**EXAMPLE:** $V_b = 8$ lpm; $C_b = 10\%$; $V_{uf} = 12$ lpm; $C_{uf} = 0.5\%$.

$$C_2 = 2.5\%. \qquad\qquad C_1 = 5.5\%.$$

**75 % Recovery**

Figure 4

## 3-STAGE RINSE

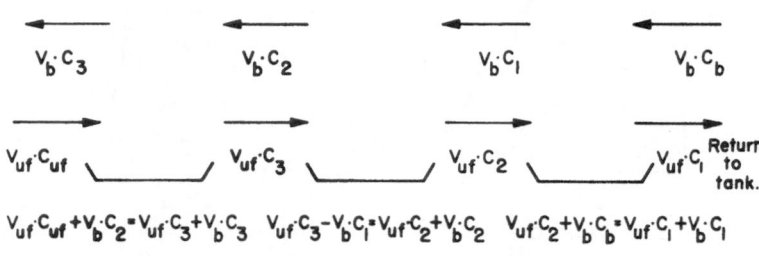

$$V_{uf} \cdot C_{uf} + V_b \cdot C_2 = V_{uf} \cdot C_3 + V_b \cdot C_3 \quad V_{uf} \cdot C_3 - V_b \cdot C_1 = V_{uf} \cdot C_2 + V_b \cdot C_2 \quad V_{uf} \cdot C_2 + V_b \cdot C_b = V_{uf} \cdot C_1 + V_b \cdot C_1$$

**EXAMPLE:** $V_b = 8$ lpm; $C_b = 10\%$; $V_{uf} = 12$ lpm; $C_{uf} = 0.5\%$.

$$C_3 = 17\% \qquad\qquad C_2 = 3.4\%. \qquad\qquad C_1 = 6.1\%.$$

**83 % Recovery**

Figure 5

BIBLIOGRAPHY

1.  Bogart, H. N., Burnside, G. L., and Brewer, G. E. F.:
    "The Concept and Development of the Ford Electrocoating
    System."  S. A. E. paper 650270 (988A), January 1965.

2.  Frangen, K. H.:  "Die Electrophorese," Farbeund Lack,
    Vol. 70, No. 4, pp. 271-279, (1964), and Vol. 12, No.
    1, pp. 36-50, (1966).

3.  Ranney, M. W., "Electrodeposition and Radiation
    Curing," Noyes Data Corp., Park Ridge, N. J. 07656,
    U. S. A., 1970.

4.  Yeates, R. L. "Electropainting," Draper Ltd., England,
    1969.

# GAS CHROMATOGRAPHIC DETERMINATION OF AMINE SOLUBILIZERS AND HYDROXYL-CONTAINING SOLVENTS IN ELECTROCOAT SYSTEMS

ELDRIGE E. WHITE and HAROLD D. SWAFFORD

GLIDDEN-DURKEE, DIVISION OF SCM CORPORATION

STRONGSVILLE, OHIO 44136

## INTRODUCTION

Commercial electrodeposition of organic coatings is a relatively new development, since as recently as eight years ago, there was only one electrodeposition primer coating tank in commercial operation in the United States.[1]  Today, the number of such installations exceeds 150 and continues to grow at a rapid rate.

The reasons for this rapid increase in popularity are not hard to find.  The process is rapid; it produces a uniformly deposited, rust and corrosion resistant coating over even welded surfaces and interior recesses of metal parts; it substantially diminishes fire hazards normally associated with coating operations; and it significantly decreases the pollution problems of conventionally applied industrial coatings.

In the electrocoating process, specially formulated paints are electrodeposited onto charged metal surfaces, applying a uniform film of paint, at a film thickness which is controlled by such factors as current, residence time in the tank, paint composition, and temperature.  Anodic coating systems usually consist of dilute solutions of alkali-solubilized polymer containing acidic functionality, together with dispersed pigments and minor quantities of various additives, which are used to help maintain stability

of the bath and to improve the properties of the electrodeposited film.

The most important additives are excess amine solubilizers and hydroxyl-containing organic solvents.  One or
more of each of these chemical types are normally included
as a part of anodic electrocoating formulations.  These
additives are commonly used at fairly low levels (from
approximately 0.1% to 1.0%), but exert a large influence on
the overall performance of the electrocoat tank system.

During the electrodeposition process, the resins and
pigments are removed from the bath more or less uniformly,
while most of the amine solubilizers and relatively high
boiling, hydroxyl-containing solvents remain behind and
tend to increase in concentration.[3,4,5]

This condition cannot be allowed to continue for very
long without encountering serious coating problems.  To
correct this situation, two basic techniques are employed
by most users of this coating process:

(1)   Excess solubilizer and coupling solvent may be
      removed from the bath on a more or less
      continuous basis, using a technique such as
      electrodialysis through a semi-permeable membrane
      ("ultrafiltration").[1,2]  Alternately, excess
      cationic bases may be removed by percolation of
      the batch through an ion-exchange column.[3]

(2)   Excess solubilizer and coupling solvents may be
      effectively removed from electrocoat baths by
      the periodic addition of replenishment feed
      material which is deficient in these additives.
      This approach provides the most economic process,
      since the excess additives are re-used, and no
      expensive treatment external to the tank is
      required.

In practice, both of these techniques are frequently
employed to varying degrees to maintain the optimum balance
between bath composition and feed material.  Regardless of
which technique is employed, it is necessary that an
analytical measurement of amine and coupling solvent
concentrations be made periodically so that precise

adjustments in component ratios can be made.

The concentration of amines present in an electrocoat bath is conventionally determined by potentiometric titration.[4-8]  By this means an amine equivalent ("MEQ" value) may be determined, which allows periodic adjustment of the total amine concentration to be made.  The disadvantage of this method is that it makes no distinction between amine types and gives no information about the relative proportions of amines present when more than one type is used.  For precise control of electrocoat bath performance, it is desirable to monitor and adjust levels of each amine type used.

The co-solvent concentration of electrocoat baths and permeates may be conveniently determined by gas chromatographic analysis of the total bath.[7]  Potential difficulties in using this method results from problems associated with finding suitable operating conditions to separate all possible combinations of volatile components likely to be encountered in electrocoat baths - particularly with unknown systems.  A flame ionization detector must be employed to eliminate interference due to the very high water content present, which adds significantly to the cost of the instrumentation required.

A gas chromatographic method has been developed which permits simultaneous qualitative and quantitative determination of essentially all of the commonly used amines and hydroxyl-containing co-solvents present in either electrocoat baths or permeates in a single analysis.  The method gives excellent precision, and permits routine analysis of a wide variety of electrocoat bath systems.

The principle of the method involves the quantitative conversion of all volatile, active hydrogen-containing compounds in the sample (amines, hydroxyl-containing solvents and water) to their corresponding trimethylsilyl derivatives while in sealed "reaction" vials, followed by gas chromatographic analysis of the derivatives, using a thermal conductivity detector.

Substitution of the silyl group, $-Si(CH_3)_3$, for active hydrogens significantly reduces the polarity of a compound and decreases the possibilities of hydrogen bonding.  The silylated derivative is usually more volatile and stable (except to water) than the parent compound.  Since 1963, this technique has been used widely by gas chromatographers to analyze many different types of high boiling and thermally unstable hydroxy and amino compounds, such as sugars, amino acids, phenols and steroids.[9]  The technique is ideally suited for analyzing the amines and hydroxyl-containing solvents used in commercial electrocoating systems.

## EXPERIMENTAL

### Gas Chromatograph

Experimental data acquired in the development of this method were obtained using a Hewlett-Packard Model 810 gas chromatograph, equipped with a Model 3370-A electronic integrator.  Any linear temperature programmable instrument, equipped with a thermal conductivity detector, may be used.

### Chromatographic Column

A 16-foot, 1/4 inch O.D. length of copper tubing is packed with 20 parts by weight of DC-11* silicone grease coated onto 60-80 mesh Chromosorb W⊛ and conditioned over-night at 300 degrees Centigrade.

### Operating Conditions

```
Detector Temperature.............. 300°C.
Detector Current.................. 150 ma
Injection Port Temperature........ 300°C.
Initial Column Temperature........ 100°C.
Final Column Temperature.......... 300°C.
Column Heating Rate............... 6°C./minute
Carrier Gas Flow Rate (Helium).... 140 cm³/minute
Carrier Gas Pressure.............. 65 psi
Chart Speed....................... 30 inches/hour
```

---

* DC-11 silicone grease is manufactured by Dow Corning Corp.

Chromosorb W is a registered trademark of Johns-Manville Products Corp.

## Reagents and Equipment

1. TMS reagent. Mix 20 parts by volume of bis(tri-methylsilyl)trifluoracetamide* with 80 parts by volume of hexamethyldisilazane. When stored in a sealed bottle, this reagent will remain stable for at least two weeks.

2. 1,4-butanediol.

3. 3 cc "Reacti-Vials"**

4. Screw cap vials of approximately 5 cc capacity

## Procedure

Weigh to the nearest 0.1 milligram, approximately 3-4 grams of electrocoat bath or permeate into a 5 cc screw cap vial. Add 20-30 milligrams of 1,4-butanediol internal standard, also weighed to the nearest 0.1 milligram, and shake thoroughly. Transfer approximately 0.1 cc by syringe to a graduated 3 cc "Reacti-Vial" and immediately add 2 to 2-1/2 cc of TMS reagent (a 20-30 fold excess should be used).

Seal the vial tightly and shake thoroughly. Place in a 60°C. oven for 2-4 hours. Reaction is complete when only one phase is observed in the vial. Withdraw 40-50 micro-liters of the reaction product by inserting a syringe through the septum top of the "Reacti-Vial". Inject into the chromatograph, using the operating conditions described above.

## Calculations

Per cent Amine equals 

$$\frac{A \times N \times D \times 100}{I \times W}$$

Per cent Solvent equals 

$$\frac{S \times N \times D \times 100}{I \times W}$$

---

* Regis Chemical Company, Chicago, Illinois.

** Pierce Chemical Company, Rockford, Illinois.

Where:  A equals  Area counts for amine peak
        I equals  Area counts for internal standard peak
        S equals  Area counts for solvent peak
        D equals  Detector response correction factor
        N equals  Weight of internal standard
        W equals  Weight of sample


## RESULTS

Table 1 lists the relative retention times and relative detector response factors for the trimethylsilyl derivatives of the amines and hydroxyl-containing solvents used in this study.


## TABLE 1

| Component | Rel.Ret.Time | Rel.Resp. |
|---|---|---|
| Cellosolve ® | 0.49 | 1.07 |
| Dimethylethanolamine | 0.52 | 1.17 |
| Monoethanolamine | 0.70 | 0.86 |
| Monoisopropanolamine | 0.75 | 1.13 |
| Diethylethanolamine | 0.77 | 1.60 |
| Butyl Cellosolve ® | 0.82 | 1.66 |
| 1,4-Butanediol (Internal Standard) | 1.00 | 1.00 |
| Hexyl Cellosolve ® | 1.18 | 1.78 |
| Diethanolamine | 1.23 | 1.08 |
| Di-isopropanolamine | 1.31 | 1.67 |
| Triethanolamine | 1.82 | 1.34 |
| Tri-isopropanolamine | 1.92 | 1.47 |

For the highest accuracy, it is recommended that users of this method determine their own relative retention times and relative response factors, using the instrumentation and exact conditions to be used in performing sample analyses.

A mixture of amines and hydroxyl-containing solvents in water were prepared at five different concentration levels to determine the absolute accuracy of the method. Results obtained are given in Table 2.

---

Cellosolve, Butyl Cellosolve, and Hexyl Cellosolve are trademarks of Union Carbide Corporation.

TABLE 2

| Component | BLEND-A | | | BLEND-B | | | BLEND-C | | | BLEND-D | | | BLEND-E | | |
|---|---|---|---|---|---|---|---|---|---|---|---|---|---|---|---|
| | % Present | % Found | Dev. | % Present | % Found | Dev. | % Present | % Found | Dev. | % Present | % Found | Dev. | % Present | % Found | Dev. |
| Monoethanolamine | 2.01 | 1.99 | -0.02 | 1.12 | 1.15 | +0.03 | 0.59 | 0.59 | 0.0 | 0.31 | 0.32 | +0.01 | 0.16 | 0.17 | +0.01 |
| Monoisopropanolamine | 1.99 | 2.01 | +0.02 | 1.11 | 1.13 | +0.02 | 0.59 | 0.58 | -0.01 | 0.30 | 0.31 | +0.01 | 0.15 | 0.16 | +0.01 |
| Diethylethanolamine | 2.00 | 2.04 | +0.04 | 1.11 | 1.11 | 0.0 | 0.59 | 0.60 | +0.01 | 0.30 | 0.30 | 0.0 | 0.15 | 0.14 | -0.01 |
| Butyl Cellosolve | 1.96 | 1.97 | +0.01 | 1.09 | 1.09 | 0.0 | 0.58 | 0.55 | -0.03 | 0.30 | 0.29 | -0.01 | 0.15 | 0.13 | -0.02 |
| Hexyl Cellosolve | 2.01 | 2.00 | -0.01 | 1.12 | 1.11 | -0.01 | 0.59 | 0.59 | 0.0 | 0.31 | 0.29 | -0.02 | 0.16 | 0.15 | -0.01 |
| Diethanolamine | 2.02 | 2.02 | 0.0 | 1.12 | 1.14 | +0.02 | 0.59 | 0.60 | +0.01 | 0.31 | 0.30 | -0.01 | 0.16 | 0.15 | -0.01 |
| Di-isopropanolamine | 2.37 | 2.38 | +0.01 | 1.32 | 1.34 | +0.02 | 0.70 | 0.70 | 0.0 | 0.36 | 0.36 | 0.0 | 0.18 | 0.18 | 0.0 |
| Triethanolamine | 2.02 | 2.02 | 0.0 | 1.13 | 1.12 | -0.01 | 0.60 | 0.60 | 0.0 | 0.31 | 0.30 | -0.01 | 0.16 | 0.16 | 0.0 |
| Tri-isopropanolamine | 2.02 | 2.02 | 0.0 | 1.12 | 1.10 | -0.02 | 0.59 | 0.60 | +0.01 | 0.31 | 0.31 | 0.0 | 0.16 | 0.16 | 0.0 |

Table 3 shows the results obtained by analyzing a typical commercial electrocoat bath (paint) and a sample of permeate (ultrafiltrate) obtained from it. The samples were run in duplicate to show the degree of repeatability that may typically be obtained using this method.

TABLE 3

### Electrocoat Bath

| Components | Run A (%) | Run B (%) |
|---|---|---|
| Monoethanolamine | 0.11 | 0.13 |
| Butyl Cellosolve | 0.70 | 0.71 |
| Diethanolamine* | 0.07 | 0.07 |
| Di-isopropanolamine | 0.64 | 0.64 |
| Triethanolamine | 0.91 | 0.93 |

### Electrocoat Permeate

| Components | Run A (%) | Run B (%) |
|---|---|---|
| Monoethanolamine | 0.04 | 0.04 |
| Butyl Cellosolve | 0.64 | 0.63 |
| Diethanolamine* | 0.08 | 0.07 |
| Di-isopropanolamine | 0.30 | 0.29 |
| Triethanolamine | 0.87 | 0.86 |

---

* Present as an impurity in the commercial triethanolamine used.

Table 4 shows the results obtained by analyzing the same electrocoat bath and permeate used for Table 3, after making known additions of the five components.

## TABLE 4

### Electrocoat Bath

| Components | % Added | % Increase Found (A) | % Increase Found (B) |
|---|---|---|---|
| Monoethanolamine | 0.83 | 0.83 | 0.80 |
| Butyl Cellosolve | 0.44 | 0.44 | 0.44 |
| Diethanolamine* | 0.06 | 0.06 | 0.06 |
| Di-isopropanolamine | 0.77 | 0.79 | 0.79 |
| Triethanolamine | 0.70 | 0.70 | 0.67 |

### Electrocoat Permeate

| Components | % Added | % Increase Found (A) | % Increase Found (B) |
|---|---|---|---|
| Monoethanolamine | 0.64 | 0.66 | 0.65 |
| Butyl Cellosolve | 0.43 | 0.43 | 0.45 |
| Diethanolamine* | 0.08 | 0.10 | 0.10 |
| Di-isopropanolamine | 0.77 | 0.79 | 0.80 |
| Triethanolamine | 0.99 | 0.98 | 0.99 |

Figure 1 shows the chromatogram obtained by analyzing a standard solution of amines and hydroxyl-containing solvents in water, using 1,4-butanediol as an internal standard. Early peaks present are due to excess reagents and the reaction product of water.

---

* Present as an impurity in the commercial triethanolamine used.

FIGURE 1

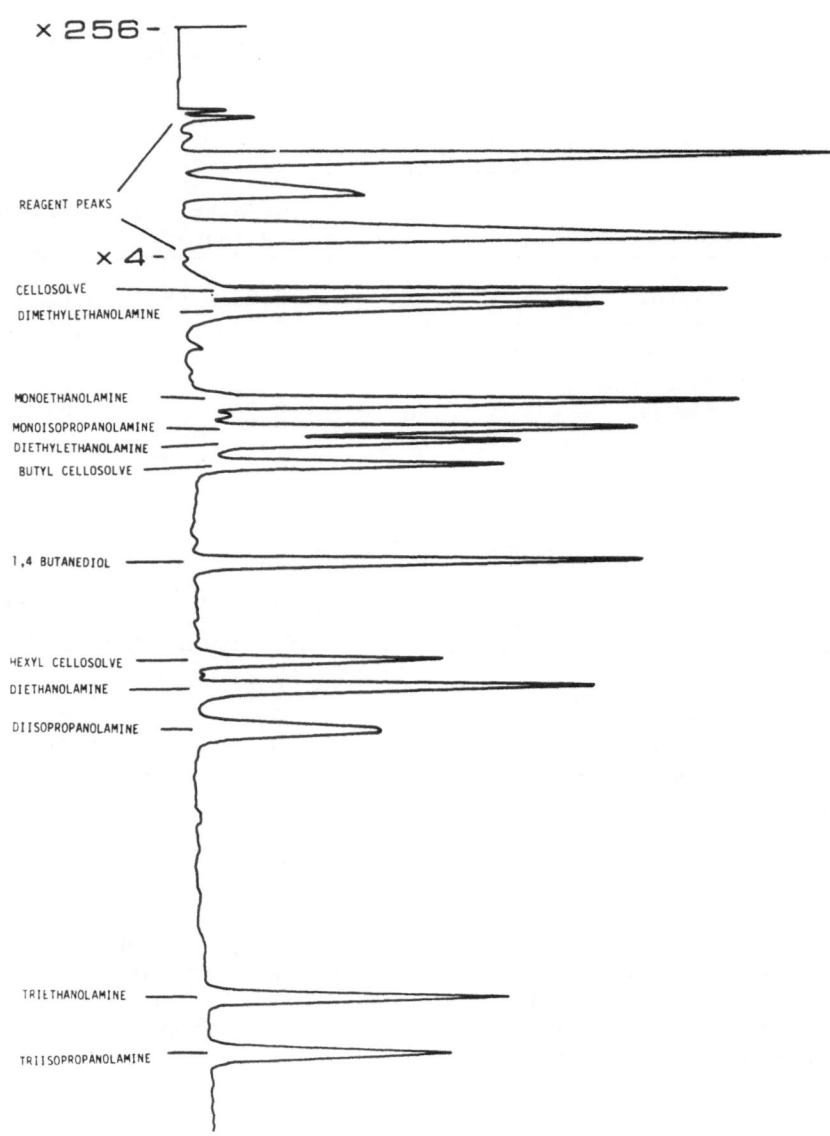

Figure 2 shows the chromatogram obtained by analyzing a commercial electrocoat bath (the same bath used for obtaining data for Table 3).

FIGURE 2

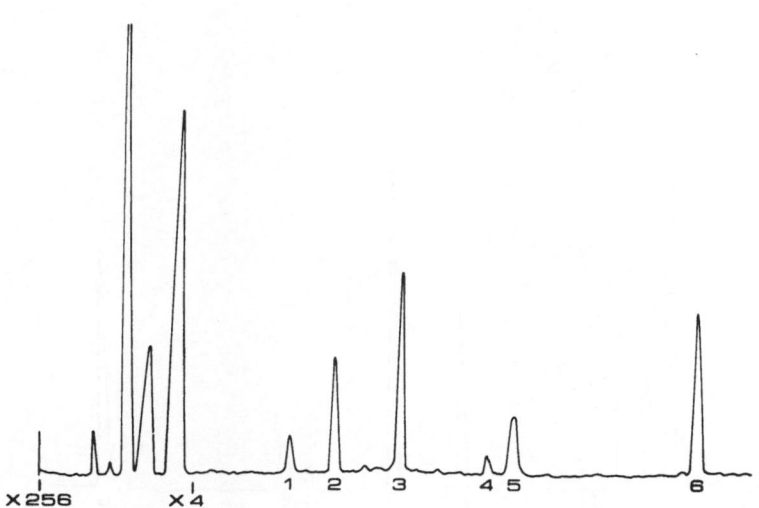

1.  Monoethanolamine
2.  Butyl Cellosolve
3.  1,4-Butanediol

4.  Diethanolamine
5.  Di-isopropanolamine
6.  Triethanolamine

Figure 3 shows the chromatogram obtained by analyzing the permeate recovered from the commercial electrocoat bath (the same permeate used for obtaining data for Table 3).

FIGURE 3

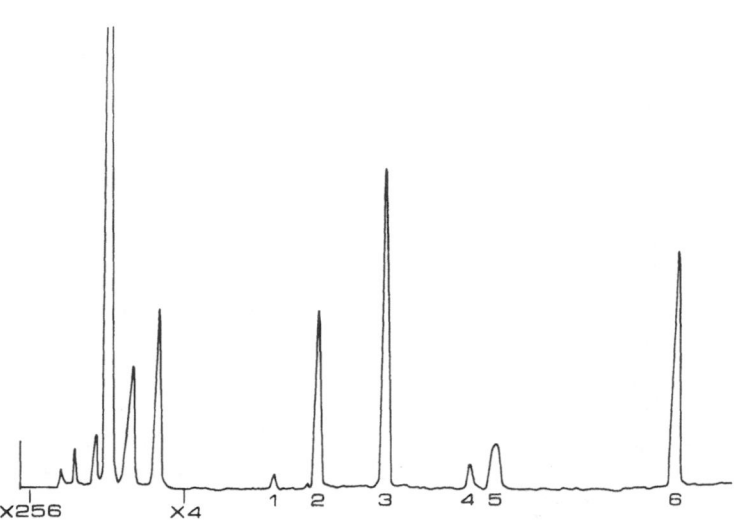

1.  Monoethanolamine
2.  Butyl Cellosolve
3.  1,4-Butanediol

4.  Diethanolamine
5.  Di-isopropanolamine
6.  Triethanolamine

## PRECISION

Standard deviations were estimated for the mean of four determinations for each of nine components at five concentrations over the range of 0.15% to 2.0%. At 0.25% concentration, the estimated standard deviation of the mean was 0.015%, indicating that the true value should be within ± 0.04% of the observed mean 95% of the time. At 1.0% concentration, the estimated standard deviation of the mean was 0.03%, indicating that the true value at this concentration should be within ± 0.07% of the observed mean 95% of the time.

## DISCUSSION

If the "Reacti-Vial" containing the sample and TMS reagent is gently agitated while in the oven at approximately 15 minute intervals, the total reaction time required can be reduced to 1-2 hours. Care must be taken, however, not to shake or agitate the vial too vigorously after it has been heated, since loss of sample can occur due to pressure build-up caused by the exothermic reaction. If time is not critical, the longer reaction time with minimum or no agitation is preferred. It is often convenient to keep the vials in the oven overnight and perform the analysis the following day, which eliminates the need for any agitation.

A standard blend of amines and solvents should be prepared, and analyzed periodically, to check column performance as well as to update relative retention time and relative response factor data.

It was experimentally established that a 20-30 fold excess of TMS reagent is required to assure quantitative conversion of all amines and hydroxyl-containing solvents to their corresponding trimethylsilyl derivatives under the conditions used in this test procedure.

The "Reacti-Vials" can be re-used indefinitely. However, the disposable septums should only be used once to prevent possible loss of volatile trimethylsilyl derivatives at elevated temperature and pressure.

Pigments present in electrocoat bath samples do not present any problems, since they readily settle to the bottom of the vial before the sample is withdrawn for analysis.

Very low boiling bases such as ammonia and triethyl-amine, as well as low boiling alcohols such as isopropanol and n-butanol, cannot be determined by this method, since their trimethylsilyl derivatives elute with the reagent peaks. A further limitation is that polyamines such as diethylenetriamine and its higher homologs cannot be determined due to the low volatility of their trimethylsilyl derivatives. Cellosolve and dimethylethanolamine may be determined by this method, but only on a qualitative basis. Accurate quantitative data are not feasible when these two components are present at a concentration below about 1%, due to the interference of minor reagent peaks which elute at the same retention time.

## SUMMARY

A gas chromatographic method has been developed which will simultaneously determine the types and amounts of most of the amines and hydroxyl-containing solvents commonly used in electrocoat bath formulations. The method is accurate, reasonably rapid, and is applicable to a wide variety of electrocoat baths and permeates. It is not applicable to the determination of low-boiling amines and solvents, or to the higher-boiling polyamines.

## REFERENCES

1.  Gerhart, H. L., *Paint and Varnish Production*, 61(2), 40 (Feb., 1971).

2.  Tanaka, T., Kusano, H., and Motoyama, Y., *J. Jap. Soc. Col. Mat.*, 43(6), 284 (1970).

3.  Burnside, G. L., Brewer, G. E. F., and Strosberg, G. D., *J. Paint Tech.*, 41 (534), 431 (1969).

4.  Motoyama, Y., Kusano, H., and Ohe, O., J. *Paint Tech.*
    41 (533), 402 (1969).

5.  Holzinger, F., *Paint Technology*, 30(8), 20 (Aug. 1966).

6.  Hezel, E., *Farbe und Lack*, 75(2), 139 (1969).

7.  Tessari, D. J. and Anderson, D. G., *Electrocoat 71:*
    A Symposium Sponsored by the NPVLA, Paper 8.

8.  Martin, C., *Paint Technology*, 33(5), 19 (1969).

9.  Pierce, Alan E., *Silylation of Organic Compounds*
    (Pierce Chemical Co., Rockford, Ill., 1968).

# THE ULTRAVIOLET CURING OF COATINGS

S.H. Schroeter

General Electric Company
Corporate Research & Development
Schenectady, New York 12345

Rapid chemical changes at or near room temperature are the basis of present industrial applications of visible or ultraviolet light in photography, photocopying and photoprinting.  The potential of using rapid photo-induced reactions that do not depend on thermal energy for the cure of coatings is attractive for three principal reasons: the steady economic drive for faster curing coatings, the increasing need to coat heat sensitive materials, such as plastics, and the mounting requirements for pollution-free processes.  In a non-thermal process, liquid, solvent-free "100 percent solids" formulations can be converted into solid coatings without loss of the  volatile low molecular weight components.  Development of light-curable formulations over the last few years has led to uv-curable systems which have found successful commercial applications in specific areas both in Europe, Japan and the U.S.  The present paper reviews some of the aspects of this new uv-curing technique.

Resins.  Present uv-curable resins are cured by photoinitiated free-radical polymerizations.  The condensation or addition reactions that are the basis of most commercial thermosetting resins such as acrylics, polyesters, melamines, silicones, epoxies and urethanes (1) cannot be carried out by ultraviolet light.  Other photocondensation reactions, although used for polymer synthesis (2), have not yet been employed for the design of coating resins.  Only in cases where the ionic condensation and addition reactions of the present resins can be achieved at room temperature through

highly active catalysts is it possible to induce such cures
by light through the photochemical generation of the catalyst.
A commercial system based on this principle (3) employs the
cure of amino resins by hydrochloric, hydrobromic or p-
toluenesulfonic acid generated photochemically from halomethyl
ketons such as p-trichloromethyl benzophenone (4,5) or p-
toluene sulfonates of $\alpha$-hydroxymethyl benzoin esters
$C_6H_5C(O)C(OR)(CH_2OSO_2C_6H_4CH_3)C_6H_5$ (6). A similar system uses
hydrochloric acid, generated by the electron beam irradiation
of carbon tetrachloride (7). Other systems, apparently still
in the experimental or developmental stage, describe ionic,
photochemically initiated cures of (a) epoxy resins and/or
vinyl ethers and oxetanes containing manganese carbonyl
compounds (8,9), polyboron compounds (10), aryldiazonium salts
(11) or derivatives of bis-fluoroalkylsulfonyl methane e.g.
$[(CF_3SO_2)_2 CH]_2Zn$ (12); (b) polymers containing N-methyl ether
units by photosensitive diazoquinones that produce hydrogen
ions in the presence of water (13); and (c) phenolic resins by
hydroiodic acid generated by the photolysis of halomethyl
benzophenones (4), halomethyl acetophenones $(CHX_2)C_6H_4C(O)$ or
$C_6H_5C(O)CHX_2$ (5) or iodoform (14). Some of these systems where
the photochemically generated catalyst does not promote the cure
at ambient temperature are useful for reprographic processes;
the latent image generated by uv via the catalyst can be developed
through subsequent heating. However such compositions may find
only limited use as coating materials.

In contrast, many free radical polymerizations proceed
readily at room temperature and are easily initiated by ultraviolet light. They are therefore the basis of current uv-
curable resins. Most of these systems consist of solutions
of unsaturated polymers dissolved in copolymerizable monomers.
Such combinations allow the formulation of liquid "100 percent
solid" resins that have acceptable viscosities at room temperature. Since no heat other than the heat of polymerization
is involved during cure, such polymer/monomer systems can be
cured without loss of the monomers in a pollution-free manner.

The unsaturated polymers used in those formulations can
be divided into two broad classes: Unsaturated polyester resins
and polymers containing pendant unsaturation.

In unsaturated polyester resins, the monomer "solvent",
primarily styrene, copolymerizes with fumarate or maleate
double bonds which are randomly distributed in the polyester

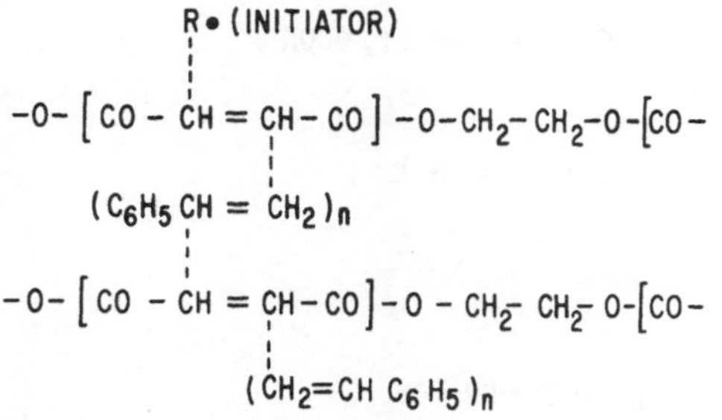

$$R\bullet(INITIATOR)$$

$$-0-\left[CO-CH=CH-CO\right]-0-CH_2-CH_2-0-\left[CO-\right.$$

$$(C_6H_5CH=CH_2)_n$$

$$-0-\left[CO-CH=CH-CO\right]-0-CH_2-CH_2-0-\left[CO-\right.$$

$$(CH_2=CH\ C_6H_5)_n$$

Fig.1  Crosslink mech-
anism in unsaturated
polyester resins.

backbone (Fig.1).   Peroxide-cured unsatuated polyesters
have been used commercially for many years (15,16).   The
resins have also been the work-horses of electron-beam
curable resins (17-28).   Much experience has therefore been
gained in these systems about cure-structure-monomer-
property relationships.   According to the patent literature,
which covers mostly the stabilizers or sensitizers discussed
below, uv-curable unsaturated polyesters are quite similar
in composition to resins used commercially for peroxide cures.
Properly uv-cured unsaturated polyesters have been found to
give properties equivalent to peroxide-cured systems.   Not
surprisingly, the uv-method therefore has found its first
application in the cure of these materials.   The process is
now used   in   furniture and cabinet manufacture where unsaturated polyesters find use as board fillers, top coats
for furniture, doors, panels, panelling, etc. (29-46).

The second major class of uv-curable polymers consists
of thermoplastic resins into which the very reactive acrylate
or methacrylate groups, or, less frequently, fumarate esters
or the less active allyl groups have been introduced.
(Isolated pendant olefinic groups as present in common alkyds
were found relatively unreactive (47); only highly conjugated
fatty acids in combinations with special dryers cured satisfactorily (48,49) and have been suggested as uv-curable
printing ink vehicles).   The reactive (meth)acrylate groups
are attached to the backbone of the resins through functional
groups such as hydroxy, carboxy, amino, isocyanato, anhydro
or epoxy by methods familiar from the design of thermosetting
resins (Fig.2).   Examples of such modified, saturated resins
which have been described include acrylics (50-58), vinyl
polymers other than acrylic resins such as styrene-allyl
alcohol copolymers (59,60), styrene-hydroxyalkylacrylate
copolymers (60,61), partially hydrolyzed polyvinylacetate (62),

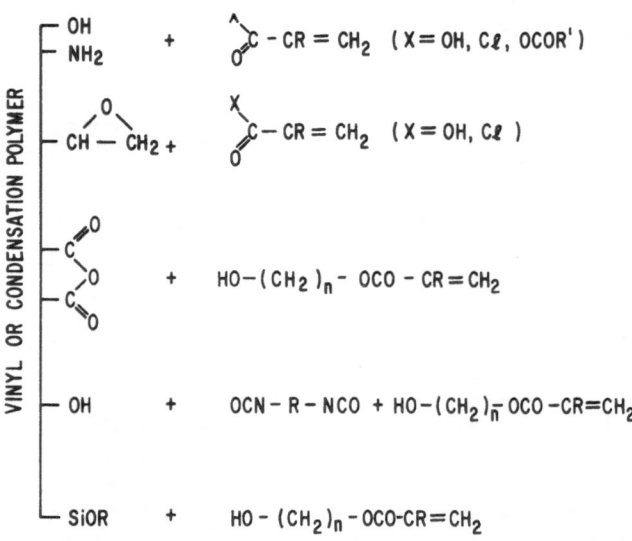

**Fig.2** Introduction of reactive groups into uv-curable
polymers.

epoxies (64-72), polyesters (73-75), polyamides (76,77),
polyurethanes (78-80), amino resins (81) and silicones (82).

Basic requirements that restrict the choice of such
modified polymers for their use in solventless compositions
are their solubility in reactive monomers and their reactivity in radical polymerizations. For example, phenolic
resins,for obvious reasons, are not suitable. The need for
solutions of reasonable viscosities usually restricts the
molecular weight of the unsaturated polymer. Solventless
compositions usually cannot simply be "thinned" at choice by
reactive monomer. For optimum cure, the ratio of monomer to
unsaturation in the polymer often has to be kept within
narrow limits (83). The choice of reactive monomers is
likewise determined by their solvent power, reactivity and
effect on resin properties. Styrene, vinyl toluene, alkyl
acrylates and methacrylates, acrylonitrile and vinyl acetate
are commonly used alone or in combination with each other.
Most of the solventless resins described can be synthesized
from common commercial monomers. Their prices should therefore be competitive with those of present thermosetting
resins. Coverage per gallon of paint will, of course, be
considerably higher for the solventless resins.

Alternate systems that have been proposed as uv-curable resins consist of liquid, high molecular weight monomers of low vapor pressure and, hence, less pollution. Most of these compounds consist of di- or higher functional acrylates derived from polyols, diisocyanates, hydroxymethylmelamine, etc. (84-93). Unless carefully formulated, coatings from such polyfunctional monomers will tend to be brittle because of the high crosslink density. They may therefore only be useful in special applications, for example, as printing ink vehicles (88-91).

Complete conversion of unsaturated double bonds in the often highly crosslinked, room-temperature cured systems calls for special care in the design and cure of all uv-curable resins. It has been shown that peroxide-cured, highly unsaturated polyesters contain considerable amounts of fumarate unsaturation in the cured resin, regardless of the amount of styrene and the method of cure (94,95). In addition, achievement of optimum properties in cured unsaturated polyesters has been linked to the complete conversion and/or removal of unreacted monomer (96,97). Studies of the electron-beam cure of mixtures containing mono and difunctional acrylates showed that the amount of difunctional monomers had to be kept within certain limits for complete conversion of all unsaturation (98,99). Surprisingly high amounts of unreacted styrene were reported to be present in certain (unspecified) uv-cured polyesters (100).

The compositions of many uv-curable-resins are quite similar to those of systems previously prepared for electron-beam curing. Pertinent references dealing with electron-beam cure are therefore cited in this article. It must be kept in mind, however, that redox-cure, electron-beam and uv-cure cannot be interchanged indiscriminately. Differences in cure and, consequently, final properties of the coatings are to be expected due to inherent differences of the cure methods; for example, the higher energy of the electrons as compared to that of the uv-light, the higher radical concentrations generated by the electron-beam, the different response of aromatic versus aliphatic molecules towards ionizing radiation, the ability of ionizing radiation to initiate both free radical _and_ ionic polymerizations (101), the greater penetration of electron-beams into organic materials, the inherently higher monomer losses during redox-initiated cures (102), etc.

Problems associated with the design and use of solventless uv-curable resin systems are, of course, in many respects
similar to those reported (103-107) previously for analogous
electron-beam or peroxide cured systems: High viscosities
resulting in unusual application properties, effect of resin
shrinkage during polymerization upon the adhesion to smooth
surfaces, air inhibition of the polymerization and potential
pollution from low boiling monomers.  Since the polymerization of styrene as well as of most acrylates and methacrylates
in inhibited by oxygen, special care must be taken to ensure
proper surface cure.

Whereas polyesters have been formulated to cure tackfree in air (15,16), the best cure of resins containing
acrylic monomers apparently requires an inert atmosphere for
cure (108).  Alternative methods used or proposed to overcome air-inhibition include the addition of phosphorus compounds (109), of stearyl stearate emulsions (110), iso-
cyanates (26,111) or drying oils (90) to the resin as well
as physical shielding by films (112,113) or the use of an
atmosphere consisting of monomer vapor and an inert gas (114).
In significant contrast to the electron-beam cure, the uv-
method allows the cure of wax-containing resins, particularly
if the somewhat slower cures by low-pressure lamps are used.

Addition of wax has also been shown to reduce evaporation losses significantly (102).  It appears that the major
pollution from uv-curing processes using solutions of unsaturated polymers in reactive monomers occurs during applications and deaeration prior to the actual cure which is
almost completely pollution-free.  Overall losses of as little
as 2 to 3 percent have been observed for properly uv-cured
wax-containing unsaturated polyester resins (102).  On the
basis of these results, the uv-cure may well be ranked among
the low polluting cure methods.  However, significant pollution problems may be expected from certain systems that use
low boiling monomers such as methyl methacrylate, acrylonitrile and so forth.  As has been pointed out (105), precautions in this area cannot be overemphasized since most
solvents are significantly toxic.

Sensitization.  Resins which undergo cure by free radical
mechanisms can be polymerized without added sensitizer by
direct exposure to intense uv-light.  However, in order to
achieve practical cure speeds and to avoid unnecessary
degradative overexposure of the resins, initiating radicals
must be produced via added uv-initiators or "uv-sensitizers".

Generation of free radicals (R·) by photo-initiators (IN) or photosensitizers (Sens) has been shown (115) to occur by two basic mechanisms:

(1)   Initiation through free radicals produced from the photochemically excited initiator IN* by a process in which the initiator itself is consumed (equations 1 and 2).

$$IN* \xrightarrow[\text{several steps}]{\text{one or}} R·  \qquad (1)$$

$$R· + \text{Resin (monomer or polymer)}$$
$$\downarrow \qquad\qquad (2)$$
$$\text{polymerization (cure)}$$

(2)   Initiation through radicals produced via energy transfer, i.e., transfer of energy from the photochemically excited sensitizer Sens* to the resin (monomer or polymer). In these processes, the initiator itself is not consumed, i.e., behaves as a true sensitizer, (equations 3,4 and 2).

$$Sens^o \xrightarrow[\text{one or more steps}]{h\nu} Sens* \qquad (3)$$

$$Sens* + \text{Resin} \longrightarrow R· + Sens^o \qquad (4)$$

In process (1), the initiator molecules may be produced by direct (unimolecular) decomposition of the electronically excited initiator, by hydrogen abstraction reaction of the photoexcited initiator (ketones, dyes, etc.) or by photo-chemically induced charge-transfer interaction between initiators (Fig.3).

For practical, commerical purposes, uv-initiators and their products must have the following properties:

(1)   Solubility in the uncured and compatibility with the cured resin systems at concentrations of about 0.1 to 5 percent by weight (or ca $10^{-2}$ mole/l).

(2)   Chemical and thermal stability in order to guarantee sufficient shelf life at ambient temperatures in the dark (usually 3 to 12 months).

HOMOLYTIC CLEAVAGE

$$AB \xrightarrow{h\nu} AB^* \longrightarrow A\bullet + B\bullet$$

$$ArCO - CR(OR') Ar, \ ArS - SAr, \ Ar-C\ell, \ R-N{=}N - R, \ RO - OR$$

HYDROGEN ABSTRACTION

$$AB \xrightarrow{h\nu} AB^* \xrightarrow{RH} ABH\bullet + R\bullet$$

$$C_6H_5COC_6H_5, \ ACRIDINE, \ DYES$$

CHARGE TRANSFER

$$AB \xrightarrow{h\nu} \left[A^+ \ B\right]^* \longrightarrow \bullet A-B\bullet$$

$$\left[Fe^{3+} \ C\ell^-\right]^* \longrightarrow Fe^{2+} + C\ell\bullet$$

Fig.3  Initiator mechanisms for free-radical photopolymer-
izations.


(3)  Acceptable physiological properties such as low
toxicity.

(4)  No negative effect on film properties.  The initiator and its photolysis products must not induce undesired
properties such as subsequent embrittlement or yellowing to
the cured films.  For use in clear lacquers, the initiator
and its products must be colorless.  Unless used in very
small concentrations, the initiator should therefore not
absorb significantly above 4000 Å.

(5)  Low cost.  Since most commercial uv-curable resins
compete with products selling for about 30 to 80¢/lb, initiator costs must be kept within definite limits.

(6)  High efficiency.  In a given resin system, the
initiator should give high cure rates, preferably at a low
initiator concentration and a low uv-intensity.  The above
described photochemical processes must therefore produce
highly effective radical chain initiators R· with high efficiency.  For a given initiator, at a given light intensity
and initiator concentration, the efficiency will depend upon
the amount of light absorbed by the initiator.  Since most
uv-curable resins show little transmission below 3000 Å, uv-
initiators that absorb only below 3000 Å will be very

inefficient in curing thicker films. Initiators must therefore absorb above 3000 Å, yet, for clear lacquers, not above
4000 Å. Ideally, the initiator should show strong absorption
both within and outside the region where the resin itself
absorbs. It must be kept in mind, of course, that the initiator must not filter out all the uv from the lower layers of
the resin. Of ultimate importance in the design of curable
resin is the determination of the most "efficient" region of
the photosensitizer. The most economical process will be
one where the "action spectrum" (i.e., the region of maximum
efficiency) of a sensitizer has been matched with the maximum uv-output of an efficient lamp.

Among the many initiators that induce free-radical
polymerizations (115-119), three major chemical classes of
compounds have been described for the uv-curing of
lacquers:

(1)  Benzoin $C_6H_5C(O)CH(OH)C_6H_5$ and derivatives presently are the most efficient sensitizers both for the cure
of polyesters and acrylic resins. Compounds of this type
include benzoin and aryl and alkyl substituted benzoins
$ArC(O)CR(OH)Ar'$ (120-122) and their ethers $ArC(O)CR(OR')Ar'$
(123-130), benzoin thioethers (123-130), $\alpha$-hydroxymethyl
benzoin and derivatives $ArC(O)C(CH_2OR)(OR)Ar'$ (133,134),
desoxybenzoins $ArC(O)CRR'Ar'$ (135), $\gamma$-benzoyl, $\gamma$-hydroxy
butyric acid and esters $ArC(O)CH(OR)CH_2COOR$ (136) and related
compounds (137). The initiators undergo $\alpha$-cleavage upon uv-
irradiation (138,Fig.3). Benzoin esters do not act as photosensitizers; they undergo intramolecular cyclization rather
than dissociation into free radicals (138). The synthesis of
specific benzoins has been covered in several patents (128-
130). Storage stability of benzoin ethers $ArC(O)CR(OR')Ar'$
depends both on the groups R and R' (125,126,130) and can
be improved by various additives (139-142).

(2)  Sulfur compounds, especially dixanthates
$RC(S)SSC(S)R$, have been described as efficient sensitizers
for acrylate resins (143-146). Aromatic disulfides $ArSSAr'$
have been claimed as efficient sensitizers for the cure of
polyesters (147-150). Although somewhat less efficient than
benzoins, they impart better pot life (thermal stability) to
the systems (141).

(3)  Halogen containing compounds, such as chlorinated
hydrocarbons (88-91,151-153), $\alpha$-halomethyl polynuclear

hydrocarbons (154-156), sulfonyl halides (157,158) and
haloketones (155,159) comprise the third class of initiators.

Phenylhydrazine has been claimed as another sensitizer
for the cure of unsaturated polyesters (160). Polynuclear
quinones (161) as well as many dye-sensitized redox systems
(162) have been described as efficient sensitizers for the
cure of acrylic monomers in printing applications. However,
no use seems yet to have been made of these systems for
coating purposes.

Studies of the uv-cure rates of resins as functions of
molecular structure, monomer content, type and concentration
of initiator, wavelength, intensity or cure temperature have
not yet been revealed. It has been mentioned that cure rates
do not increase linearly with light intensity and that care
must therefore be taken in projecting commercial cure rates
from laboratory action spectrum measurements (163). Kinetic
data obtained previously from the photopolymerization of
monomers in solution will only partly be applicable to a
small range of the bulk-polymerization processes, where the
system changes rapidly from low to extremely high viscosities
(95,164).

The relative low absorption of the presently used initiators in the region of the maximum uv-output of most lamps
allows the cure of resin films of 100 mil or more, e.g. of
laminates or films (165-167). The uv-cure of resins in the
presence of large amounts of translucent fillers such as talc,
asbestos, calcium carbonate or barium sulfate presents the
largest present industrial application of uv-curing. Films
pigmented with $TiO_2$ are not cured by uv (34); successful
cures with dyes and selected pigments have been reported (41).
More research will be needed to define the limits of curing
pigmented systems by ultraviolet.

Systems. Standard uv-lamps, suitable for uv-curing, are
commercially available from lamp manufacturers. Both medium
and high pressure mercury arcs as well as special low pressure mercury lamps coated with phosphors ("black light" or
"actinic lamps") are employed commercially as uv-sources
(34c,35,163,168), others such as Xenon flash lamps (169,170)
and plasma arcs (171) have been proposed. Medium pressure
mercury lamps at their present practical maximum loadings of
about 200 watts/inch produce about fifty times the uv-intensity
of actinic-lamps (163). Mercury arcs are therefore the sources

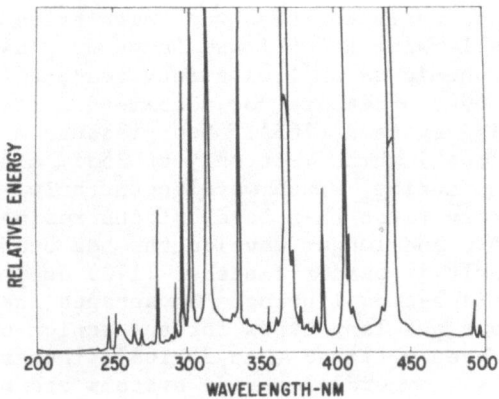

**Fig.4** Spectral energy distribution of medium pressure
mercury arcs.

of choice where maximum cure speeds are desired.  They do,
however, offer the design problem of adequate heat removal.
Medium pressure lamps emit over the whole ultraviolet with a
fixed (relative) spectral distribution (Fig.4).  The spectral
power distribution of black lamps on the other hand can be
varied in the near ultraviolet by proper choice of phosphors
(Fig.5).  This offers the possibility of matching maximum
lamp output with the "action spectrum" (i.e., the region of
maximum "efficiency") of the sensitizer.

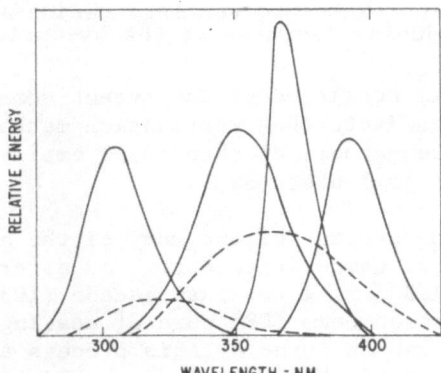

**Fig.5** Spectral energy distribution of black light fluorescent
lamps.

Still higher uv-intensities than with present mercury arcs are available with Xenon flash lamps or plasma arcs. However, for economic as well as safety reasons (violent failure) these uv-sources are not recommended for use in commercial curing systems (163). Low pressure mercury lamps ("germicidal lamps") which emit only at 2537Å alone are not useful for paint curing, since wavelengths below 3000Å only penetrate the very first thin layer of the resins. A combination of short and longer wavelengths has been reported, however, to result in harder finishes (172) due, probably, to enhanced surface-crosslinking. Advantages and disadvantages of various lamp-types for the design of uv-curing equipment have been outlined (34c,35,163). Integrated pilot plants as well as commercial curing systems are now being offered by a number of suppliers in Europe and the U.S. Designs that allow cures of wide areas at the lowest possible temperatures have been described (30,173,174) Present commercial resin/system combinations using medium or high pressure mercury arcs allow cure speeds of 15 seconds or less (175). Line speeds used in industrial installations vary from about 10 to 100 ft/min.

<u>Cost Estimates</u>. Application of uv-curable finishes has been considered as an alternative to solvent-based thermosets, heat-cured polyesters, laminates or electron-beam-cured coatings.

Low capital investment for uv-systems, reduced labor cost and reduced factory space (due partly to the higher cure speeds), savings in pollution equipment, and higher coverage per gallon of the solventless resins (ca. 1,500 $ft^2/$ mil) are cost-reducing features of the uv-curing process.

Calculations, confirmed by the recent commercialization of the process, indicate that fabrication methods using uv-curing can be more economical than those employing thermally cured coatings or laminates (39,44).

The uv-curing process offers many of the advantages of radiation curing at competitive cost. An alternative process, extensively studied during the last decade (103-107,176-183) involves the electron-beam (EB) cure of coatings. Highly pigmented resins can be cured by this process at speeds even higher than those available by the uv-method. Commercially, because of high investment costs, this process is not competitive with thermal finishing operations unless used for

extremely large operations. It has therefore only recently
found its first industrial use as the "Electrocure Process"
in coating car panels (184,185) and in coil coating (186).
Estimates of the economics of EB curing vary widely (25,39,
106,187-194), depending on the assumptions made on equipment
cost, required dose, film thickness, efficiency of energy
use, cost of auxiliary equipment such as screening and inert
gas generator, capacity, and number of sides to be coated.
It seems safe to assume that at the present time, the EB
process, as quoted,becomes competitive in finish operations
that coat at least 20,000,000 $ft^2$/year (193-194).

UV and EB-curing therefore supplement each other in
many respects. The uv-method may also be used more advantageously, for example, in the cure of shaped articles since
uv-lamps can easily be arranged to surround any given configuration. It may also be used where substrates such as
paper or certain plastics, which are degraded or discolored
by electron-beam radiation are to be cured. It may further
be employed advantageously for the cure of resins such as
unsaturated polyesters which usually require uneconomically
high doses for electron-beam cure. On the other hand, the
EB-method offers advantages in speed and substrate penetration. It has been suggested (39) that, due to high electrical
costs for present uv-systems, UV and EB-cure may become equally
economic attractive at certain production volumes.

Presently, commercial uv-curing applications are used
primarily in the furniture area. There is ample evidence,
however, that the technique will also find use in the can,
electrical, plastic, paper and printing industry. Undoubtedly, both UV and EB-curing will find their place in other
coating operations. Predictions are that as much as 20 percent of all industrial finishes may be cured by some form of
radiation in the U.S. by the end of the decade (195-197).
Significant cross-fertilization is to be expected between EB
and UV curing technology in the design of coating compositions.
Undoubtedly, the uv-cure of coatings will also benefit from
knowledge gained in the commercializiation of uv-curable
printing inks (198-201), printing plates (202), and photo-
reprographic photopolymer systems (203).

## REFERENCES

1. P. Nylen and E. Sunderland, "Modern Surface Coatings", Interscience, New York, 1965

2. J.G. Higgins and D.A. McCombs, Chemtech., March 1972, p.176

3. Anon, Farbe u. Lack, 77, 348 (1971)

4. Netherland 70.13779; Mar.22, 1971; to Bayer

5. Netherland 70.14179; Mar.22, 1971; to Bayer

6. R. Heine, H. Rudolph and H.-J. Kreuder; Ger. Offen. 1,919,678; Nov.5, 1970; to Bayer

7. US 3,522,159; July 28, 1970; to Kansai Paint Co.

8. W. Strohmeyer and P. Hartmann, Z. f. Naturforschung, 24b, 939 (1969)

9. W.S. Anderson, J. Appl. Polym. Sci, 15(8), 2063 (1971)

10a. Belg. 644,590; Sept.2, 1964; to DuPont

10b. W.E. Mochel, US 3,196,098; July 20, 1965; to DuPont

11. Netherland 755,013; Aug.19, 1970; to American Can Co.

12. Fr. Demande 2,038,085; Jan.8, 1971; to 3M Co.

13. W. Laessig, H. Ulrich, E. Mueller, and K. Dinges, US 3,533,796; Oct.13, 1970; to Gevaert-Agfa

14. J.L. Silver; German 1,772,274; Oct.28, 1971; to Union Carbide Corp.

15. H.V. Boenig, "Unsaturated Polyesters", Elsevier Publ. Co., New York, 1964

16. W.A. Riese, "Loesefreie Anstrichsysteme", C.R. Vincentz Verlag, Hannover 1967

17a. J.V. Schmitz and E.J. Lawton, US 2,900,277; Aug.18,1959; to General Electric

17b. E.J. Lawton, US 2,997,419; Aug.22,1961; to General
     Electric

18.  British 828,717; Febr.24,1960; 907,688; Mar.24,1960;
     and 949,191; Febr.12,1964; to T.I (Group Services) Ltd

19.  US 3,247,012; April 19,1966; US 3,437,513; and
     3,437,514; both April 8,1969; to Ford Motor Co.

20.  J.R. Guenther and R.B. Mesrobian, US 3,246,054; April 12,
     1966; to Continental Can Co.

21.  G. Zeppenfeld et al., E. German 77,067; Nov.12,1970;
     and 78,374; Dec.12,1970

22.  US 3,619,392; Nov.9, 1972; to Bayer

23.  W.J. Burlant and J.E. Hinsch, J. Polym. Sci (A), $\underline{2}$, 2135
     (1964); ibid. $\underline{3}$, 3587 (1965).

24.  J.D. Nordstrom, 4th Int'l. Conf. on Electron and Ion Beam
     Science and Technol., 1970, Sect.IV, p.605

25.  F.L. Dalton, Plastics and Polymers, Oct.1970, p.343

26.  G.J. Pietsch, ACS, Div. Org. Coatgs and Plastics, $\underline{29}$(1),
     1969, Minneapolis Meetg

27.  S.I. Omel'chenko, N.G. Videnina, V.G. Matjushova,
     I.N. Chervetsova, and G.N. Pyankov; Ind. Eng. Chem. Prod.
     Res. Develop., $\underline{9}$(2), 143 (1970)

28.  A.S. Hoffman, J.T. Jameson, W.A. Salmon, D.E. Smith,
     and D.A. Tragesei, ibid., $\underline{9}$(2), 158 (1970)

29.  E.W. Rabehl, Ind.-Lack.-Betr.; $\underline{36}$, 147 (1968)

30.  E. Wiederhold, Am.Paint J., Dec.9,1968, p.68

31.  Anon, Arbeitskreis Holz, Mering, $\underline{85}$, 3 (1968)

32.  Anon, Ind.-Lack.-Betr., $\underline{36}$, 479 (1968)

33.  Anon, Paint Technol, Aug.1968, p.35

34.  W. Deninger and M. Patheiger, Farbe u. Lack, 74, 1129
     (1968); ibid., 75, 893 (1969); Ind. Lack Betr., 37, 84
     (1969); J. Oil Col. Chem. Assoc., 52, 930 (1969);
     Holztechnol., 11(1), 20 (1970); Paint e Varnish Prod.,
     Febr.1970, p.44

35.  M. Patheiger, Oberflaeche-Surface, 10(7), 472 (1969)

36.  Can. Paint e Finish, March 1969, p.73

37.  Anon, Ind. Finish, Sept.1969), p.49

38.  Anon, Woodworking e Furniture Dig., Sept.1970, p.53

39.  P.G. DeLange, Verfkroniek, 43, 105 (1970)

40.  R. Hindley, Ind. Finish, 22, April 1970, p.29

41.  Anon, Ind. Finish, Nov.1969, p.49; Nov.1970, p.44

42.  W.E. Harris, Ind. Finish., 46, Nov.1970, p.44

43.  R.A. Helmers, Furniture Design e Manuf., May 1971, p.33;
     ibid., Aug.1971, p.40

44.  R.L. Koch, Am. Paint J., July 26,1971; p.10

45.  W. Brushwell, Am. Pt.J., Aug.1971, p.57

46.  R.L. Koch, Ind. Finish, 47(8), 44 (1971)

47a. R. Klose, Ind.-Lack.-Betr., 36(6), 234 (1968)

47b. J. Oliver, Prod. Finish, April 1969, p.74

48.  A. Bassl and K.-H. Boehm-Kasper, Farbe e Lack, 73(10),
     916 (1967)

49.  German 1,794,230; Oct.28,1971; and Netherland 70.06363
     Nov.5,1970; both to Bayer

50.  French 1,489,458; Nov.13,1967; to Ford (France)

51.  A.C. Schoenthaler, US 3,418,295, Dec.24,1968; to DuPont

52. J.D. Nordstrom and J.E. Hinsch, Ind. Eng. Chem. Prod. Res. Develop., 9(2), 155 (1970)

53. Brit. 1,955,271, May 27,1970, to Kansai Paint Co.

54. US 3,530,100; Aug.9,1970; to PPG

55. W.J. Burlant, US 3,437,514; April 8,1969; to Ford

56. W.J. Burlant and C.R. Taylor, US 3,528,844; Sept.15, 1970; to Ford

57. G.F. D'Alelio, US 3,557,063; Jan.19,1971 to PPG

58. G.F. D'Alelio, US 3,654,251; April 4, 1972; to PPG

59. Netherland 69.10077; Jan.5,1970; to Ford

60. S.B. Radlove and A. Ravve, US 3,546,002, Dec.8,1970; to Continental Can Co.

61. Netherland 69.10079; Jan.5, 1970; to Ford

62. Netherland 69.10078; Jan.5,1970; to Ford

63. T. Kimura et al., Ger. Offen 1,964,547; July 23,1970; to Mitsubishi Rayon Co. Ltd.

64. E.J. Aronoff et al., US 3,586,526; June 22, 1971; to Ford

65. E.J. Aronoff and S.S. Labana, US 3,586,527; June 22, 1971; to Ford

66. S.S. Labana, US 3,586,528; June 22, 1971; to Ford

67. E.J. Aronoff, S.S. Labana, and E.O. McLaughlin, US 3,586,529; June 22, 1971; to Ford

68. E.J. Aronoff, S.S. Labana, and E.O. McLaughlin, US 3,586,530, June 22, 1971; to Ford

69. S.S. Labana, Polym. Prepr., 10(1), 502 (1969); and refs.98,99

70. L.S. Miller, US 3,560,237; Feb.2,1971; to Weyerhaeuser Co.

71. Ger. Offen 2,013,471; Oct.15,1970; to Dainippon Ink Chem.

72. M.F. Blin, G. Gaussens, and F. Lemaire, Ger. Offen
    2,123,092; Nov.25,1971; to Comm. a l'Energie Atomique

73. G.F. D'Alelio, US 3,485,732, Dec.23,1969; to PPG

74. Netherland, 69.18533; June 16,1970; to Bayer

75. Netherland, 70.1328; Aug.4,1970; to Com. à l'Energie
    Atom.

76. H. Takeshi and H. Kodama; French 2,057,914; June 25,1971

77. German 1,917,798; Oct.15,1970; to PPG

78. French 1,513,285; April 30,1968; to Ford (France)

79. German 1,916,499; Nov.13,1969; to Weyerhauser Co.

80. T.J. Miranda; US 3,600,359; Aug.17,1971; to O'Brien Corp.

81. US 3,535,148; Oct.20,1970; to Continental Can Co.

82. Netherland, 69.17326; May 20,1970; to Ford

83. R.L.E. Van Gasse, US 3,562,125; Feb.9,1971; to N.V.
    Chem. Ind. Synres

84. British 1,256,859; Dec.15,1971; to PPG Ind.

85. C.W. Fitko, US 3,567,494, Mar.2,1971; to Continental
    Can Co.

86. Belg 772,251, June 9,1971; to Kalle AG

87. E.D. Feit, Polymer Preprints, 13(1), 474 (1972)

88. R.W. Bassemir, D.J. Carlick, and G.E. Sprenger,
    US 3,551,246; Dec.29,1970; to Sun Chem. Corp.

89. G.I. Nass, R.W. Bassemir, and D.J. Carlick,
    US 3,551,311; Dec.29,1970; to Sun Chem. Corp.

90. R.W. Bassemir, R. Dennis, and G.I. Nass,
    US 3,551,235; Dec.29,1970; to Sun Chem. Corp.

91a. R.W. Bassemir, D.J. Carlick, and G. Feig,
     US 3,558,387, Jan.26,1971; to Sun Chem. Corp.

91b. Ger. Offen. 2,026,311; Jan.28,1971; to PPG

92.  Netherland 69.18462; June 12,1970; to Ciba

93.  "Patents dealing with the UV or EB initiated polymer-
     ization of acrylic compounds," Sartomer Resins,
     Essington, Pa.

94.  K. Hamann, W. Funke, and H. Gilch, Angew. Chem., 71,
     596 (1959)

95.  K. Horie, I. Mita, and H. Kambe, J. Polym. Sci (A-1),
     8, 2839 (1970)

96.  B. Alt, Kunststoffe 54, 738 (1964)

97.  J. Fritz, ibid., 47, 710 (1957)

98.  S. S. Labana, J. Polym. Sci. (A-1), 8, 179 (1970)

99.  S.S. Labana and E.J. Aronoff, J. Paint Techn., 43, 77
     (1971)

100. H. Klug and M. Velic, Farbe u. Lack, 77(3), 231 (1971)

101. A. Chapiro, "Radiation Chemistry of Polymeric Systems",
     Interscience, 1962

102. S.H. Schroeter and J.E. Moore, following article

103. T.J. Miranda and T.F. Huemmer, J. Paint Technol., 41,
     118 (1969)

104. W.H.T. Davison, J. Oil Col. Chem. Assoc., 52, 946 (1969)

105. Ntl Coil Coaters Assn, Panel Presentation and Discussion
     of EB Curing, Oct.8,1969; Chicago, Ill.

106. G. Gaussens, Peint.-Pigm.-Vernis, 46(12), 1457 (1970)

107. A.S. Hoffmann, At.En. Rev., 9(2), 347 (1971)

108. T.F. Huemmer and J.R. Ungurait, Abstracts, 4th Intl. Conf. on Electron and Ion Beam Sci.e Technol., 629

109. C.H. Krauch and H. Hoffmann, Ger. Offen. 2,012,221; Sept.23, 1971; to BASF

110. Ger. Offen. 2,038,883; Febr.18,1971; to H. Wiederhold

111. E. Langholf and H. Behling, E. Ger. Appl. 77,770; Nov.20,1970

112. Netherland 67.09072; Jan.2,1968; to Ford

113. R.P. Hall, US 3,644,161; Feb.22,1972; to SCM Corp.

114. L.S. Miller, US 3,520,714; July 14, 1970; to Weyerhaeuser Co.

115. G. Oster and N.L. Yang, Chem. Rev., 68, 125 (1968)

116. J.F. Rabek, Photochem. e Photobiol., 7, 5 (1968)

117. J. Kosar, "Light-Sensitive Systems", John Wiley and Sons, New York, 1965

118. J.J. Bulloff, Bull Reg. Tech. Conf. Soc. Plast. Eng., Mid-Hudson Sect., Oct.15-16, 1970, p.73

119. J.B. Rust, ibid., p.55

120. C.L. Agre, US 2,367,661, Jan.23,1945; to DuPont

121. R.E. Christ, US 2,367,670; Jan.23,1945; to DuPont

122. M.M. Renfrew, US 2,448,828; Sept.7,1948; to DuPont

123. J.L. Crandall, US 2,722,512; Nov.1,1955; to DuPont

124. H.-G. Heine, K. Fuhr, H. Rudolph, and H. Schnell, US 3,607,693, Sept.21, 1971; to Bayer

125. K. Fuhr, R. Rudolph, H. Schnell, and M. Patheiger; US 3,639,321; Feb.1,1972; to Bayer

126. K. Fuhr, H. Rudolph, and W. Metzner, US 3,582,487; June 1,1971; to Bayer

127. French 2,029,779; to BASF

128. Netherland 70.00525; July 20,1970; to Bayer AG

129. Netherland 71.03993; Sept.28,1971; to Kon.Ind.Mij.
     Noury e Van der Lande NV

130. Netherland 69.16459; Mar.11,1970; to Bayer

131. C.C. Petropolos, J. Polym. Sci. (A), $\underline{2}$, 69 (1964)

132. French Demande 2,006,309; Dec.26,1969; to Bayer AG

133. H.-G. Heine, K. Fuhr, H. Rudolph, and H. Schnell;
     US 3,657,088; April 18,1972; to Bayer AG

134. H.Hoffmann, H. Hartmann, and C.H. Krauch, Ger. Offen
     1,923,266; Nov.19,1970; to BASF

135a. French 1,450,589; Nov.11,1966; to Muanyagipari Kutato
      Intezet.

135b. Netherland Appl. 298,715; Aug.10,1965; to Deutsche
      Gold u. Silber Scheideanstalt

136. Netherland 71.06311; Nov.10,1971; to Bayer AG

137. Belg 753,154; Dec.12,1970; to AKZO, N.V.

138. J.C. Sheehan and R.M. Wilson, J. Am. Chem. Soc., $\underline{86}$,
     5277 (1964)

139a. B. Passalenti, S. Vargiu, and U. Nistri;
      US 3,616,366; Oct.26,1971; to Soc. Ital. Resine S.p.A.

139b. U. Nistri, S. Vargiu, B. Passalenti, and O. Fiorani,
      US 3,627,657; Dec.14,1971; to Soc. Ital. Resine S.p.A.

140. Netherland 71.09084-85; Jan.4,1972; to Sta Italiana
     Resine S.p.A.

141. Ger. Offen 1,945,725; Mar.11,1971; to Chem. Werke Huels

142. Netherland 71.01198; Aug.9,1971; to Reichhold-Albert-
     Chemie AG

143.   H.L. Gerhart; US 2,673,151; Mar.23,1954; to PPG

144.   V.A. Engelhardt and M.L. Peterson, US 2,716,633;
       Aug.30,1955; to DuPont

145.   German 1,037,132; August 21, 1958; to Rohm e Haas

146.   German 1,813,011; July 9,1970; to BASF

147.   L.M. Richards, US 2,460,105; Jan.25,1949; to DuPont

148.   H. Rudolph, K.H. Mueller, and H. Schnell, **W. German**
       1,233,594; March 28,1968; to Bayer AG

149.   P. Brown, Metal Finish., Febr.1970, p.53

150.   M. Patheiger, Oberflaeche-Surface, $\underline{10}$(7), 472 (1969)

151.   Z. Reyes, Ger. Offen. 2,062,563; Sept.2,1971; to
       McCall Corp.

152.   P.G. Garratt, Ger. Offen. 2,025,789; Dec.23,1970;
       to Lonza AG

153.   M. Marx and A. Zosel, US 3,627,659; Dec.14,1971; to
       BASF

154.   C.C. Sachs and J. Bond, US 2,505,067 and 2,505,068;
       April 25,1950; to A.H. Kerr and Co.

155.   C.C. Sachs and J. Bond, US 2,548,685; April 10,1951;
       to A.H. Kerr and Co.

156.   M.C. Brodie, US 3,326,710; June 20, 1967; to Sherwin
       Williams Co.

157a.  C.C. Sachs and J. Bond, US 2,579,095; Dec.18,1951;
       to A.H. Kerr and Co.

157b.  B.J. Sites and M.S. Agruss; US 3,052,568; Sept.4,1962;
       to Miehle-Goss-Dexter

158a.  W. German 2,043,352; April 29,1971; to Vianova
       Kunstharz AG

158b.  German 1,949,010; April 1, 1971; to Bayer

159.  German 1,949,010; April 1,1971; to Bayer

160.  Japan 70.24,227; August 13,1970; to Tokyo Shibaura
      Electric Co.; see CA.<u>74</u>, 88,715y  (1971)

161.  US 2,951,758; Sept.6,1960; to DuPont

162.  US 3,558,322; Jan.26,1971; to DuPont and earlier patents

163.  R.E. Levin and H.H. Homer, Ann. Techn. Conf. Illum. Eng.
      Soc., Aug.17-20,1971, Chicago, Ill., p.111

164.  K. Demmler, Farbe u. Lack, <u>75</u>, 1051 (1959)

165.  P. Borrei and J. Lehureau, US 3,655,483; Apr.11,1972

166.  P. Borrell and J.L. Lehureau, Ger. Offen 2,059,588;
      June 24,1971, to PROGIL S.A.

167.  V.I. Horn and R.L. Novkov, Plast. Technol., April 1966,
      p.56; W.M.H. Moore and B.H. Trachtman, Sampe J., Aug/
      Sept 1968, p.9

168.  Anon, Paint e. Varnish Prod., Febr.1970, p.47

169.  B.L. Sites and M.S. Agruss, US 3,008,242; Nov.14,1961;
      to Miehle-Goss-Dexter Inc.

170.  A.C. Keyl and M. Brodie, US 3,511,687, May 12,1970; to
      Sherwin-Williams Co.

171.  C.L. Osborn and D.J. Trecker, Ger. Offen 2,003,132,
      July 30,1970, to Union Carbide Corp.

172.  Ger. Offen. 1,802,904, Febr.5,1970; to H. Wiederhold

173.  Brevet Belge 757.249 and 757.250; to S.P.R.L. Euripe

174.  G. Hestermann, Ger. Offen. 1,953,074, Oct.29,1970; to
      Eisenmann KG

175.  Am. Paint J., June 14,1971; p.9; Mtls Eng., Sept.1971,
      p.46; Paint e Varnish Prod., Febr.1972, p.43

176.  A. Bertrand, Rev. L'Inst. Francois Du Petrole, <u>24</u>, 1151
      (1969)

177.  W.B. Pederson and K. Singer, Facsim. Rep., U.S. Atom.
      En. Comm., Oak Ridge (1969)

178.  H. Wittcoff, Paintindia, Aug.1970, p.35

179.  G. Zeppenfeld, G. Zeppenfeld, and W. Ebert, Holztechnol.,
      11, 13 (1970)

180.  W. Brushwell, Am. Paint J., Aug.1971, p.65

181.  M.F. Valero, Double Liaison, 18, No.188,151 (1971)

182.  D.R. Bailey, Paint Techn., 35(10), 23 (1971)

183.  S. Young, S. Taylor, and K.W.G. Butcher, Prod. Finish.,
      23(6), 36 (1970)

184.  A.S. Turner, 4th Intl. Conf. on Electron and Ion Beam
      Sci. e. Technol., 619 (1971)

185.  Paint e Varnish Prod., April 1971, p.35

186.  Anon, Ind. Finish; April 1970, p.85; ibid., Febr.1971,
      p.31

187.  L. Kuehn, Ind.-Lack.-Betr., 35(6), 233 (1967)

188.  W. Brenner and W.F. Oliver, Reinforced Plastics,
      June 1967, p.294

189.  K.H. Morgenstern, Ind.-Lack.-Betr., 36(9), 376 (1968)

190.  E. Rotkirch, Fiber Board Coating Economy, IAEA-Symposium,
      Munich 1969

191.  H. Duerr and K. Gaefgen, Ind.-Lack.-Betr., 37(7), 285
      (1969)

192.  Anon, Farbe e. Lack; 75(6), 583 (1969)

193.  J.E. Hinsch a. C.J. Sajewski; Paint, Oil e Colour J.,
      155, June 13, 1969, p.1033

194.  S.V. Nablo, Proceedings of the 1971-Ann. Mtg. of the
      Ntl.Coil Coaters Ass., Las Vegas, Nev., p.7

195.  J.A. Mock, Matls. Eng., 56 (1970)

196.  M. Henertz, Ind. Finish., June 1971, p.7

197.  Paint e Varnish Prod., Nov.1971, p.28

198.  Chem.e Eng.News, Jan.25,1971, p.35; ibid., Oct.1971, p.38

199.  G. Nass, American Ink Maker, Jan.1971, p.25; Anon, ibid., March 1971, p.24

200.  Chem. Week, March 8, 1972, p.44

201.  Design News, May 8, 1972, p.60

202.  Chem e Eng. News, Aug.26,1968, p.62

203.  "Reprography" in Kirk-Othmer Encycl. of Chem. Technol. 17, p.328, John Wiley e Sons, New York, 1968

# THE ULTRAVIOLET CURE OF SOLVENTLESS RESINS--
# A POLLUTION FREE METHOD?

S. H. Schroeter and J. E. Moore

General Electric Company
Corporate Research and Development
Schenectady, New York 12345

## INTRODUCTION

A potential way of overcoming the air pollution problem associated with solvent-based coatings exists in the use of so-called solventless, 100 percent solids coatings. One common variety of solventless resins consists of solutions of organic prepolymers containing aliphatic unsaturation dissolved in copolymerizable organic monomers. Examples of such resins are solutions of unsaturated polyesters in styrene or vinyl toluene, or solutions of polymers containing pendant esters of acrylic or methacrylic acid in acrylate or methacrylate monomers.[1] Since no solvent has to be evaporated from such systems prior to cure, little if any pollution might be expected from such resins, particularly if such compositions could be cured in very short times at low temperatures. UV and electron beam curing have been considered as cold energy sources for this purpose.[1,2]

The present paper summarizes investigations to determine the degree of pollution to be expected from UV-curable solventless resin systems consisting of solutions of prepolymers in co-polymerizable monomers. In most of the present

work, a general-purpose unsaturated polyester
was used which contained 30 percent by weight of
various monomers. Solutions of other resins in
the same monomers may be expected to show similar
behavior.

EXPERIMENTAL - RESULTS AND DISCUSSION

In general, four sources of air pollution
were considered:

(1)   Pollution occurring during
      application (rolling, spraying,
      etc.).

(2)   Pollution prior to cure, due
      to evaporation.

(3)   Pollution during cure.

(4)   Pollution after cure, for example,
      from heating or postbaking of the
      UV-cured resins.

Application Losses

Typical monomer losses during application
of thin films were determined by drawing 2 x 2
inch films on glass slides from 30 percent monomer
solutions with a four mil draw-down blade and
weighing the films immediately after application.
The weight losses occurring after evaporating
the monomers from these films by heating for 1
hour at 110°C, subtracted from the original
monomer content, gave the following application
losses (%):   styrene, 1.4; vinyl toluene, 0.5; t-
butylstyrene, 0; acrylonitrile, 15; methyl methacrylate, 5.9; ethyl acrylate, 8.4; butyl acrylate,
2.1. Although the method of application is somewhat arbitrary, it is obvious that the use of
volatile monomers such as ethyl acrylate or methyl
methacrylate in UV-curable resins may lead to
considerable pollution. Spray application of
systems containing certain monomers in prepolymers
may not be acceptable in many instances.

## Evaporation Losses

Evaporation losses were determined by following evaporation from 2 x 2 inch films of various defined thicknesses on glass slides in air at ambient temperatures as illustrated in Fig. 1. Results from these experiments can be summarized as follows:

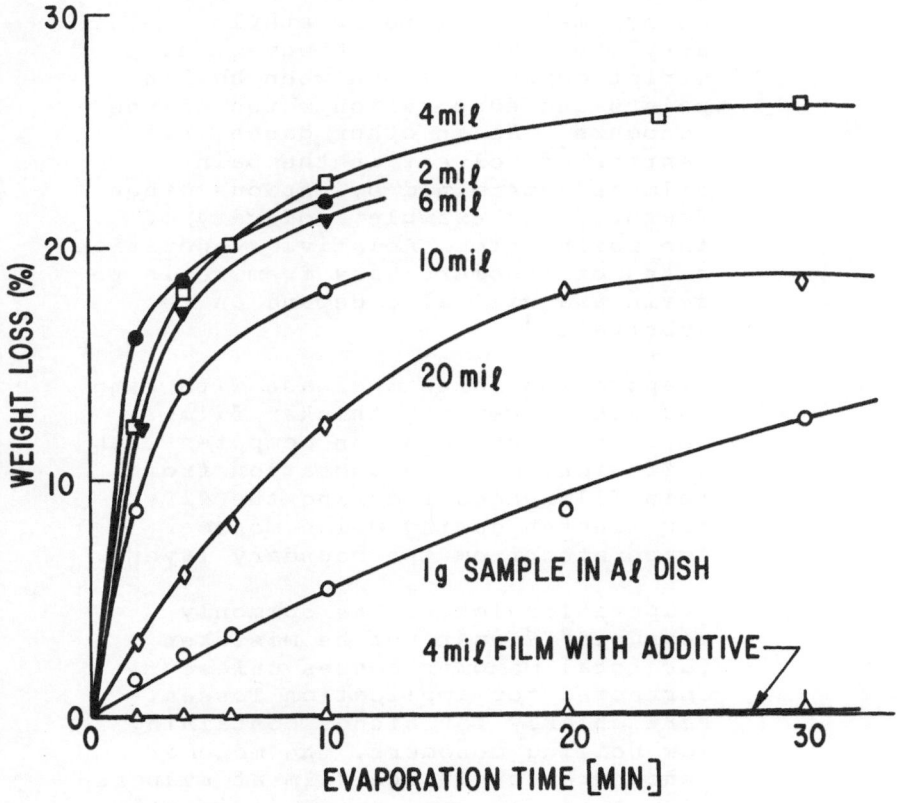

Figure 1

Room temperature evaporation losses of unsaturated polyester films from 33 ± 3% styrene solutions.

(a)     Thin resin films of about 2 to 6 mils
        containing ca. 30 percent of monomers
        evaporate substantial amounts of
        monomer at room temperature in air,
        as much as 1/3 of the total monomer
        within two minutes.

(b)     In general, evaporation rates are
        slower for higher boiling monomers,
        such as vinyl toluene, than for
        lower boiling compounds, such as
        methyl methacrylate or ethyl
        acrylate.  There is, however, no
        strict correlation between boiling
        points and evaporation rates of the
        monomers.  As in other cases, re-
        tention of solvent in the paint
        film is determined by various other
        factors, for example, polarity of
        the resin, etc.  Relative evaporation
        rates of monomers vary from resin to
        resin and will also depend on the
        substrate.[3]

(c)     Evaporation rates decrease with time
        and are slower for thicker films.
        Obviously, evaporation competes with
        diffusion; major evaporation from
        thin films occurs during the first
        few minutes during which monomer
        evaporates from the boundary layers.

(d)     Evaporation losses, as commonly
        measured[4-8] must not be mistaken
        for total monomer losses unless
        corrected for application losses.
        Particularly for blends containing
        low boiling monomers, the monomer
        concentration in the film at evapora-
        tion time zero may be significantly
        lower than in the original solution.
        Thus, artificially low evaporation
        rates may be observed.  (See Figure 2
        "corrected" versus "uncorrected"
        evaporation curves for acrylonitrile.)

The presence of certain waxes in the polyester-monomer blends was found to inhibit evaporation from thin films almost completely (Figs. 1 & 2). This was found to be the case not only in polyesters containing styrene as noted previously,[4-8] but surprisingly was also found for other monomers such as acrylates and methacrylates. Many waxy materials were tried as evaporation inhibitors such as Chlorowax, Napcowax, Acrawax, Epolene wax, carnuba wax, beeswax, ceresine wax, Oxazoline wax, stearic acid, stearyl methacrylate, methyl stearate, paraffin waxes and mineral oil. The most effective additives were the straight chain hydrocarbon paraffin waxes. Addition of talc or other fillers did not reduce the rate of evaporation.

The wax-containing resins also gave smaller application losses than the ones without wax. For example, 3 vs. 6 percent loss for methyl methacrylate and 1 vs. 15 percent for acrylonitrile (Fig. 2).

Most of the previous literature describing the use of wax in polyester films to obtain air drying properties is based on redox initiation of the polymerization. The surfaces produced tend to be waxy with poor gloss. They have to be sanded and polished to be commercially acceptable. This problem may be peculiar to the redox initiated systems. We also obtained hazy films by the redox initiated curing of resin formulations which gave clear, glossy films when properly cured by UV.

Total monomer losses, including both application and evaporation losses, as occurring prior to UV-cure from 4 mil wet films, are presented in Fig. 2 for various compositions. Major differences between monomers appear during application and the early stages of evaporation where diffusion is not yet rate determining.

### Cure Losses

Monomer losses during UV-cure were determined either by curing wax-free resin systems and subtracting

evaporation losses prior to cure from the total
observed losses or by curing wax-containing resins

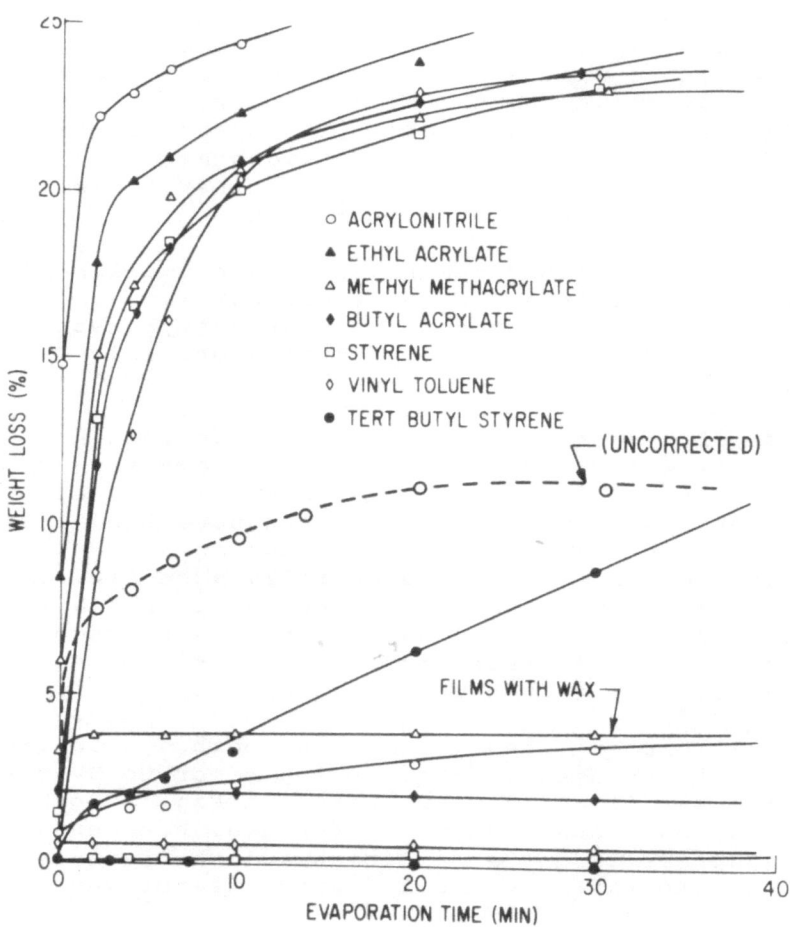

Figure 2

Room temperature application and evaporation
losses of 4 mil thick unsaturated polyester films
from 31 ± 2% monomer solutions.

that did not show any evaporation losses prior to
cure.   Cures were carried out at room temperature
in air with the unfiltered light of medium pressure
mercury arcs operated at ca. 200 watts/inch.   Cures
must be conducted at low temperatures with good heat
removal in order to minimize losses.   Whenever the
infrared was not filtered from the lamps, monomer
losses were considerably higher.   Essentially no
cure losses were observed for fast curing 4 mil
wax-containing films cured at room temperature on
substrates in close contact with a cold surface.
If the resins cure slowly, as much as 5-10% of
the monomer can evaporate from the heat of the lamp
during the cure.   The combined weight losses from
both evaporation and UV-curing of polyester films
containing various monomers as a function of film
thickness are reported in Figure 3.   Thicker films
(20 mils) both wax-free and wax-containing, which
were cured at room temperature with insufficient
cooling, showed losses of 1.5 + 0.5%, probably due
to the exotherm during polymerization.

Post-Cure Losses (Baking Losses)

Pollution may also result from under-cured
films containing unreacted monomer, specifically
where cured resins are used at higher temperatures,
for example, as electrical insulations or encap-
sulations.   Most properly UV-cured unsaturated
polyesters containing styrene and similar monomers
experience weight losses in the order of 0.3 to
3.0 percent, usually less than 1 percent, when
heated for one hour at 150°C in air.   Unreacted
monomer, thermal degradation, and/or the presence
of volatiles in the base resin are responsible
for these losses.

Bake losses from UV-cured solventless resins
are a convenient measure of cure.   This method of
determining unreacted monomer agrees well with
results from the direct chemical analysis[9] and also
with measurements of physical properties.   The
bake method has successfully been used by the
authors to follow the UV-cure as a function of sensitizers, fillers, inhibitors, light intensity,
time, etc.   See Figure 4.

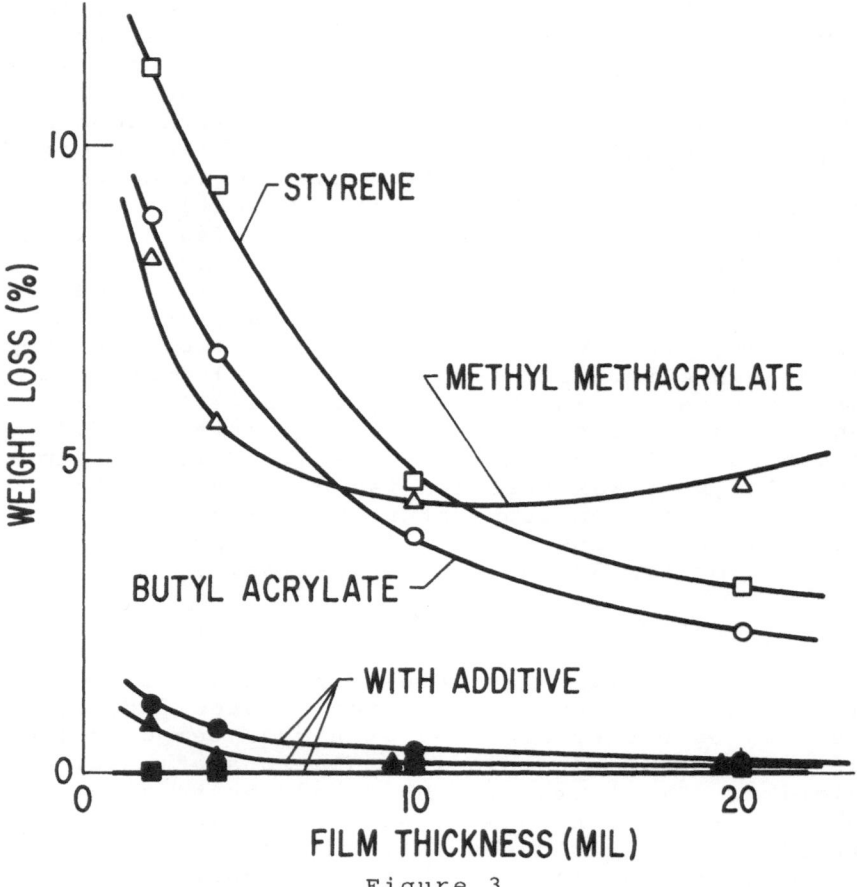

Figure 3

Room temperature evaporation and UV-cure losses
from unsaturated polyester films from 31 ± monomer
solutions.

### Thermal Cure

Thermal cures of a wax-containing polyester
formulation containing various peroxide initiators
were carried out at various temperatures using both
one gram samples in aluminum dishes and 4 mil films
on glass slides.  The results in Table 1 show that

the monomer losses for the thicker one gram samples are considerable, while those for the 4 mil films are total.   When the cures were followed for various times for two formulations at 50°C (Table 2), it became apparent that even at short cure times at low temperatures most of the monomer evaporates before polymerization.   The data in Table 3 show

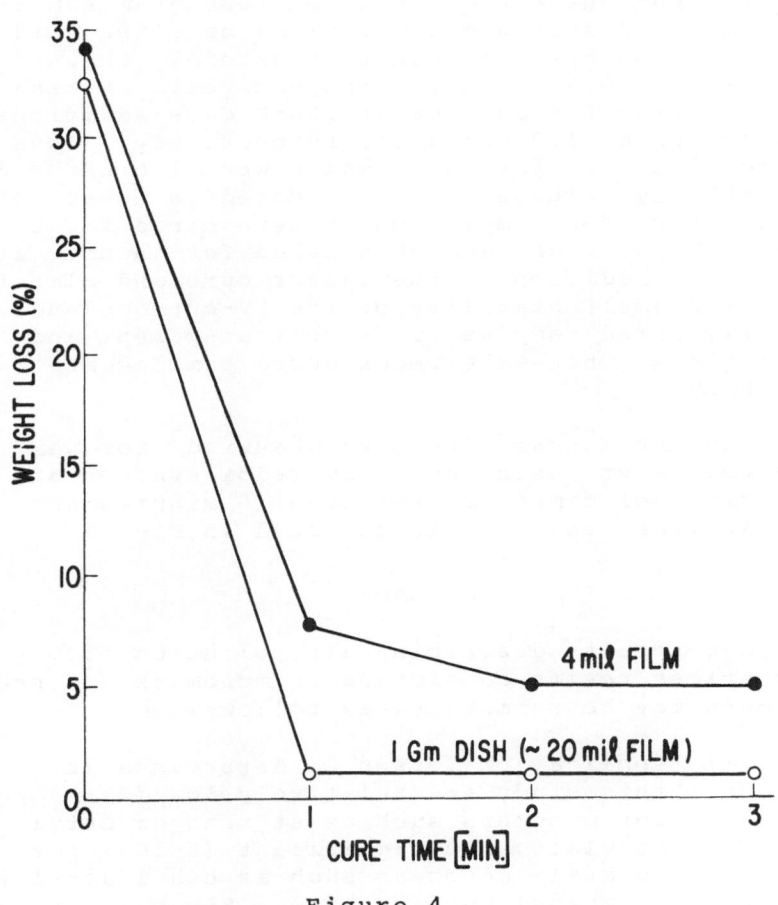

Figure 4

Weight losses during 1 hour bake at 150°C for polyester samples UV-cured for various times.

that vinyl toluene is just as volatile as styrene
whereas t-butyl styrene shows reduced volatility.
Unfortunately, the t-butyl styrene solutions tend
to be so viscous that they are difficult to use.

### Redox Cure

The same wax-containing polyester which had
been used for the UV and thermal cure study was
also cured by the methyl ethyl ketone peroxide-
cobalt naphthenate redox system. One gram samples
in aluminum dishes and 4 mil films on glass slides
were cured 60 hrs. at room temperature. These
samples lost 0.3 and 1.3%, respectively. Baking
these samples for an hour at 150°C gave additional
bake losses of 1.2 and 4.5%, respectively. Thus,
the total losses for the samples were 1.5 and 5.8%,
respectively. These may be compared to losses of
1.5 and 5.0% for samples which were cured for 2
minutes by UV light and then baked for an hour at
150°C. In addition to the faster cure and almost
unlimited shelf stability of the UV-cured blends,
properly cured samples are also transparent and
quite glossy while the redox cured samples are
very hazy.

Monomer losses have been measured[7] for wax
free polyester resins cured by redox systems at
20°C with gel times greater than 10 minutes and
are identical to those illustrated in Fig. 1.

### SUMMARY

Our results describing air pollution from
solventless resins consisting of monomers and prepolymers may be summarized as follows:

(1)   Application losses as determined in
      this study are relatively low (2%)
      for monomers such as styrene or butyl
      acrylate, but significant (5-10%) for
      volatile monomers such as ethyl acrylate
      or methyl methacrylate. Spray applica-
      tions must be considered carefully.

Table 1

Thermal Cure of Wax-Containing Polyester (ca. 33% Styrene) Using Various Peroxide Initiators

| Peroxide | Benzoyl Peroxide | | Methyl Ethyl Ketone Peroxide | | t-Butyl Perbenzoate | | Dicumyl Peroxide | |
|---|---|---|---|---|---|---|---|---|
| Sample | (a) | (b) | (a) | (b) | (a) | (b) | (a) | (b) |
| **75°C Cure** | | | | | | | | |
| Peroxide Half Life | 10 Hrs. | | 160 Hrs. | | 460 Hrs. | | 1000 Hrs. | |
| Cure Loss % (c) | 30.7 | 22.6 | 31.4 | 21.6 | 31.5 | 24.4 | 30.7 | 24.6 |
| Bake Loss % (d) | 2.7 | 0.6 | 3.6 | 1.0 | 3.1 | 2.6 | 3.3 | 3.7 |
| Total Loss % | 33.4 | 23.2 | 35.0 | 22.6 | 34.6 | 27.0 | 34.0 | 28.3 |
| **100°C Cure** | | | | | | | | |
| Peroxide Half Life | 1/2 Hrs. | | 15 Hrs. | | 18 Hrs. | | 90 Hrs. | |
| Cure Loss % (c) | 31.1 | 21.3 | 30.4 | 22.6 | 32.4 | 28.5 | 32.2 | 29.0 |
| Bake Loss % (d) | 2.7 | 0.4 | 2.5 | 0.9 | 2.1 | 0.9 | 2.3 | 1.3 |
| Total Loss % | 33.8 | 21.7 | 32.9 | 23.5 | 34.5 | 29.4 | 34.5 | 30.3 |
| **125°C Cure** | | | | | | | | |
| Peroxide Half Life | 1-1/2 Min. | | 1.6 Hrs. | | 48 Min. | | 27 Hrs. | |
| Cure Loss % (c) | 28.9 | 19.6 | 28.6 | 20.3 | 32.8 | 26.0 | 31.7 | 27.6 |
| Bake Loss % (d) | 2.0 | 0.4 | 1.2 | 0.6 | 1.6 | 0.5 | 2.4 | 0.6 |
| Total Loss % | 30.9 | 20.0 | 29.8 | 20.9 | 34.4 | 26.5 | 34.1 | 28.2 |

(a)  4 mil film.
(b)  1 gm dish.
(c)  One hour cure.
(d)  One hour bake at 150°C.

Table 2

Thermal Cure of Wax-Containing Polyester (ca. 33%
Styrene) Using Lauroyl and Benzoyl Peroxides as
Initiators at 50°C

| Peroxide | Lauroyl Peroxide | | Benzoyl Peroxide | |
|---|---|---|---|---|
| Peroxide Half Life | 48 Hrs. | | 190 Hrs. | |
| Sample | (a) | (b) | (a) | (b) |
| **10 Minute Cure** | | | | |
| Cure Loss % | 21.6 | 14.8 | 21.1 | 15.3 |
| Bake Loss% (c) | 10.1 | 6.1 | 10.8 | 5.3 |
| Total Loss % | 31.7 | 20.9 | 31.9 | 20.6 |
| **30 Minute Cure** | | | | |
| Cure Loss % | 24.4 | 17.4 | 24.0 | 17.7 |
| Bake Loss % (c) | 7.5 | 4.1 | 8.3 | 3.7 |
| Total Loss | 31.9 | 21.5 | 32.3 | 21.4 |
| **1 Hour Cure** | | | | |
| Cure Loss % | 26.9 | 19.4 | 25.4 | 19.0 |
| Bake Loss % (c) | 5.7 | 3.0 | 7.0 | 2.8 |
| Total Loss % | 32.6 | 22.4 | 32.4 | 21.8 |

(a)    4 mil film.
(b)    1 gm dish.
(c)    1 hour at 150°C.


Table 3

Thermal Cure of a Polyester Containing ca. 30%
Monomer

| Monomer | Styrene | Vinyl Toluene | t-Butyl Styrene |
|---|---|---|---|
| Cure Loss % | 20.4 | 21.1 | 10.9 |
| Bake Loss % (1/2 hr.) | 0 | 0.1 | 0.1 |
| Total Loss % | 20.4 | 21.2 | 11.0 |

Cure - 1/2 Hour at 150°C, 1 Gm Samples, 1% t-Butyl
Perbenzoate Initiator

(2)   Evaporation of most monomers from
      thin paint films is significant at
      room temperature without additives.

(3)   Evaporation of most monomers from
      paint films can be minimized by the
      addition of small amounts of wax to
      the resin blends.

(4)   For wax-containing films containing
      monomers such as styrene, which are
      cured at low temperatures, overall
      losses resulting from application,
      deaeration, cure, and subsequent
      baking (1 hr. at 150°C), may be as
      low as 2-5% by total weight.

## REFERENCES

1.   S.H. Schroeter, preceding abstract.

2.   W. Deninger and M. Patheiger, Farbe & Lack,
     74, 1179 (1968); ibid., 75, 893 (1969).
     J. Oil Col. Chem. Assoc., 52, 930 (1969);
     Holztechnol., 11, 1 (1970). T.J. Miranda and
     T.F. Huemmer, J. Paint Technol., 41, 118 (1969).

3.   W. Gebhardt, W. Hermann and K. Hamann,
     Farbe & Lack, 64, 303 (1958).

4.   B. Parkyn and E. Bader, Brit. Pat. 713,332,
     Aug. 11, 1954.

5.   T.L. Phillips, Brit. Plast., Feb. 1961, p. 69.

6.   F.V. Jenkins, A. Mott, and R.J. Ricker,
     J.O.C.C.A., 44 (1), 42 (1961).
     (See Reference 8, p. 336.)

7.   Double Liaison (Paris) 74, 25-28 (Oct. 1961).
     (See Reference 8, p. 344.)

8.   W.A. Riese, Loserfreie Anstrichsysteme,
     Curt R. Vincentz, Verlog Hannover 1967.

9.   W. Funke, H. Roth and K. Hamann, Kunststoffe,
     51, 75 (1961).

SOME FUNDAMENTAL ASPECTS OF THE POLYMERIZATION OF VINYL
MONOMERS WITH ELECTRON BEAMS

E. P. Stahel, C. C. Allen, D. R. Squire,
V. Stannett
North Carolina State University
Department of Chemical Engineering
Raleigh, North Carolina

## INTRODUCTION

The radiation induced polymerization of vinyl monomers
has been intensively studied during the past twenty-five years.
In general the polymerization has been found to proceed via
free radical mechanisms although more recent work shows that
ionic mechanisms are also operative under certain conditions[1].
Most of the research has been conducted at moderate dose rates,
up to one megarad per hour, for example, using cobalt 60 or
other convenient radioisotopes. The high penetrating power
and low dose rates minimized the experimental difficulties involved in such research. The development of high power electron accelerators in the 200-500 Kv range for high speed industrial radiation processing has led to enormous interest in
their use including the polymerization and crosslinking of solvent free organic coatings. Fundamental research, on the other
hand, using accelerators for polymerization studies has been
rare. This has been due to the experimental difficulties
caused by (a) the low penetrating power and (b) the heat effects arising from the rapid deposition of energy and the exo-
therms arising from the rapid polymerization reactions. The
penetration pattern leads to difficulties with uniform fields,
accurate dosimetry and to difficulties of conducting experiments
in an oxygen free atmosphere together with other control problems. The kinetics of polymerization themselves are further
complicated by the use of rapidly scanned electron beams rather
than a steady uniform field.

These various difficulties will be discussed briefly in
this paper together with some results obtained under carefully
controlled conditions.


HEAT EFFECTS

When monomer is irradiated, there will be heat generated
from any exothermic polymerization which occurs and from any
radiation absorption.  This heat generation may either heat
the monomer or be transferred to some adjacent region of lower
temperature.  Both of these processes occur simultaneously in
systems of interest at high dose rates.

Mathematical models have been derived and solved numerically.    These enable prediction of the gross heat effects
for the different geometries of the cylindrical ampoule and
semi-infinite plane sheet.  Essentially the approach was to
model the ambient heat exchange at the boundary, the dissipation through the substrate or wall if any, and the temperature
distribution throughout the reacting media.  These three differential equations were solved simultaneously by numerical
methods as there is no practical approach for an analytical
solution.  The assumptions made are thought to be acceptable
in light of the time scale of events at high dose rates and
are the typical ones of constancy of properties including the
heat transfer coefficient at the boundary.  Due to the speed
of such electron accelerator processes natural convection within the monomer was not considered and the kinetics developed
later in this paper were assumed to be obeyed.

An analytic solution was derived in Appendix 1 for situations where the reaction exotherm was not significantly alter-
ed by the internal film temperatures to significantly change
the overall exotherm, that is, uniform heating.  Figure 1 is
a dimensionless plot of the results for the planar geometry.
It enables rapid estimation of the maximum temperature in the
film being cured.  For the general case of a substrate, film,
and ambient gas envelope, there is a non-symmetrical temperature distribution.  Solutions for the model can be presented
as profiles through the film.  Figure 2 shows the effect of
changing film thickness on the profile.  The shape of the profile is, of course, governed by the relative temperatures of
the substrate and the ambient gas.  Cooling of the substrate
can result in the maximum shifting towards the ambient interface and occuring in the top 10% of the film.

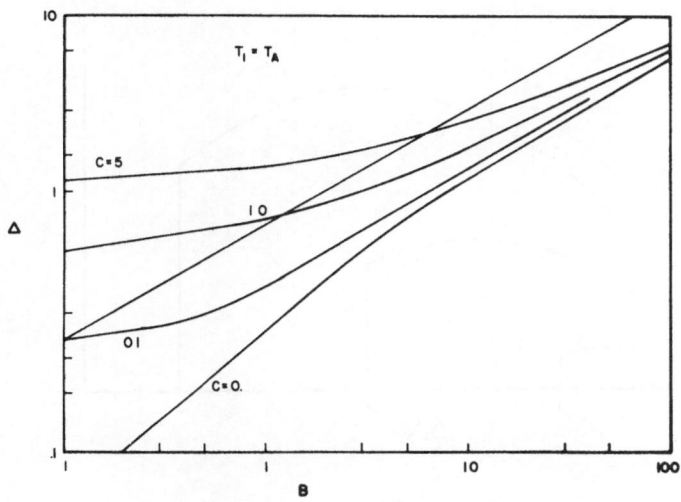

Figure 1.   Estimation of Maximum Film Temperatures

$$\Delta = -1 + \sqrt{1 + \frac{2}{\psi A^2}(T_M - T_A)} + \frac{2}{\psi A^2}(T_M - T_1) = \frac{X K_A}{K}$$

$$C = \frac{2K_A^2}{\psi K^2}(T_M - T_1)$$     $K_A$     Overall Heat Transfer Coef (Air)

     $K$     Thermal Conductivity (Film)

$$B = \frac{2K_A^2}{\psi K^2}(T_M - T_A)$$     $\psi$     Heat Generation/unit volume/K

     $X$     Film thickness

$$D = \frac{K_A X}{K}$$     $T_A$     Ambient Temperature

     $T_1$     Substrate temperature

$$A = \frac{K}{K_A}$$     $T_M$     Maximum Temperature

Beam heating can be massive, 11°C per megarad for materials like aluminum and carbontetrachloride.  The use of these
heat transfer calculations are severalfold.  These include estimation of maximum temperature changes, changes in polymerization
rates, effects on the degree of polymerization, and the reduc-

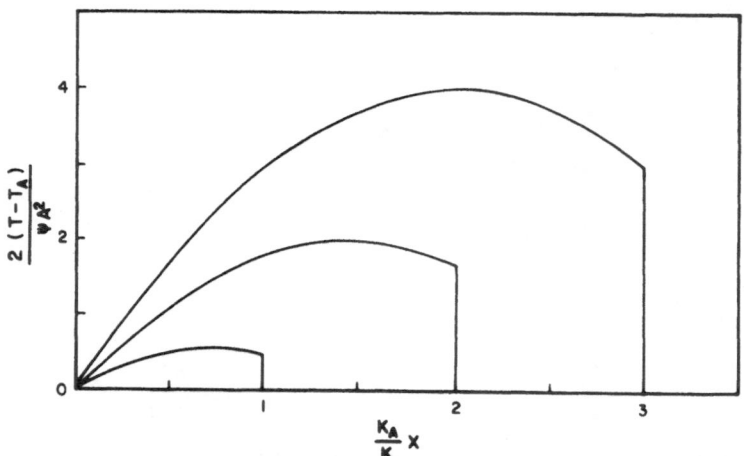

Figure 2.   Effect of Film Thickness on Temperature
            Distributions


tion of experimental data to equivalent isothermal behavior.

Temperatures in an irradiated film will increase until
they attain their "steady state" profiles.  Figure 3 shows the
effect of time on the temperature profile of a methyl methacrylate film being cured on an isothermal substrate.  These results were obtained by a six-point implicit numerical simulation
of the defining differential equations with the following assumptions:  isotropic heating in the film, constant temperature
substrate  and the heat flux at surface proportional to the temperature drop across a vapor blanket.

The temperature distribution in the film may be characterized by a time constant, which is the length of time required
for a position in the film to attain 63.2% of its final change
in temperature "steady state."  This time constant is the maximum temperature change in the film divided by the rate of change
in temperature at the onset of irradiation.

$$\tau = \frac{\Delta T_{MAX}}{\frac{\partial T}{\partial t}/t=0}$$

The time constant may be used to estimate the transient temperature response with the following equation:

$$\frac{T-T_o}{T_F-T_o} = 1 - e^{-\frac{t}{\tau}}$$

$T_F$ is the "steady state" temperature at a position in the film and T is the temperature at that position at time t.

Since the maximum temperature in the film has been shown to be proportional to the rate of heat generation, and the initial rate of increase in temperature is also proportional to the rate of heat generation, the time constant is independent of the rate of heat generation.

$$\tau = \frac{\Delta T_{MAX}}{\psi K} \rho C_p$$

The time constant would be more sensitive to changes in film thickness than conductivities, heat transfer coefficients, densities, and heat capacities.

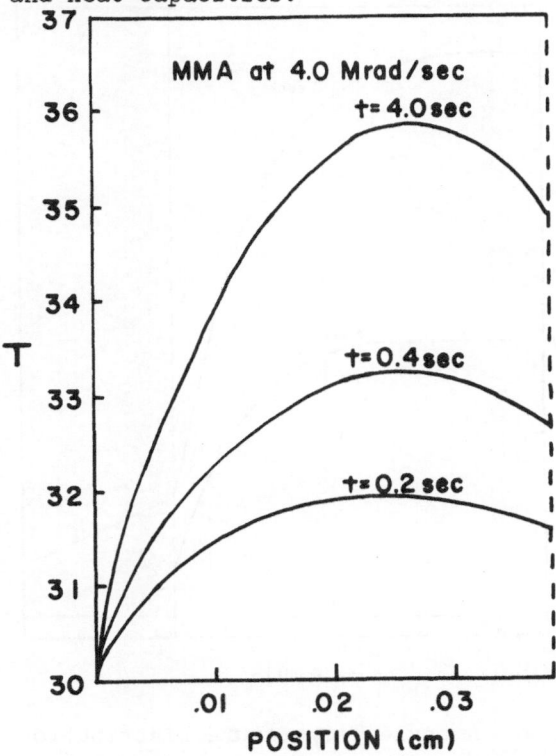

Figure 3.   The Effect of Time Constant on Temperature Distribution.

The solutions presented also can be used for ampoule studies indicating the inadequacy of ice water cooling or the
like.   Figure 4 shows three profiles at different times for
a Mrad/sec. irradiation of a typical organic monomer.   Close
to adiabatic behavior is seen for even long irradiation times
and any estimation by steady-state heat transfer approaches
would be irrelevent.   The diffusion of oxygen can be considered analogous to the heat transfer by conduction and inhibition
profiles generated.

If an ampoule containing methyl methacrylate is irradiated
for one second at 3 Mrad/sec. absorbed dose for water, pyrex
ampoule, and MMA with no heat transfer, the temperature of the
monomer would increase 14.7°C as the temperature of the water
and ampoule increased 6.9°C and 45°C, respectively.   The heat
from the ampoule is not expected to be as influential as one
would first expect since a vacuum insulates the surface of the

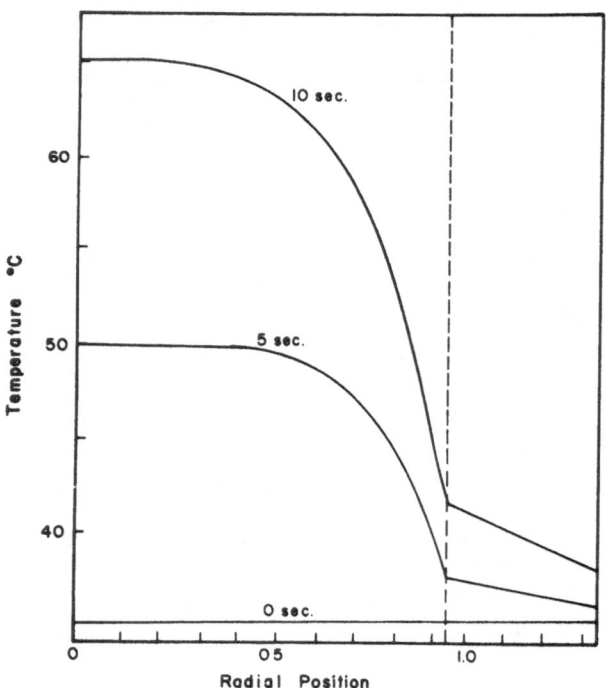

Figure 4.   Ampoule Temperature Distributions

liquid in the horizontal ampoule from the top of the ampoule and the electron beam is expected to be highly attenuated at the bottom of the ampoule. Since MMA reacts at 1%/sec. at 3 Mrad/sec. (30°C), there would be a temperature increase of

$$\frac{(1\%/sec.)(9.4 \text{ moles/liter}) \; 13.8 \text{ Kcal/mole}}{939 \text{ gm/liter} \; 0.49 \text{ cal/gm°C}} = 2.8°C/sec.$$

due to the reaction. The temperature of methyl methacrylate rises 18°C in only one second under these conditions.

The difficulty of separation of the attendant heat transfer and chemical kinetic effects in high dose rate studies need not generate the absolute pessimism heretofore shown, it is felt by the authors. Careful limiting studies both experimentally and model wise can aid in reducing the problem to a tractable kinetic level. The added benefit of such approaches can be seen in understanding the seemly anomolous data that abounds in a growing literature. Such an example is the absurdly high activation energy of 35 Kcal/gmole encountered by German workers[2] beam polymerizing acrylonitrile in a continuous system. Data such as this can be shown to include a massive heat transfer effect.

### SOME HIGH DOSE RATE KINETIC CONSIDERATIONS

Most kinetic aspects of radiation induced reactions examined in great detail in the last 20 years are discussed by Chapiro[3] in his excellent book. The absence of fundamental data at the high dose rates possible with electron accelerators has left this kinetic region a matter of conjecture. Recently, a rather complete study[4] of styrene at up to 3 Mrad/sec. has shown definitely that ionic polymerization is an important mechanism. No similar evidence has been obtained for methyl methacrylate in a parallel study. Further an equimolar mixture of styrene and methyl methacrylate yielded a "copolymer" very high in styrene content; thin layer chromatography however showed it to consist of polystyrene together with some equimolar copolymer. This presumably reflects the concurrent cationic and free radical processes operating. The kinetic aspects of the ionic polymerization at high dose rate are presented in the former paper and will not be discussed further here. Careful examination of the free radical homopolymerization of methyl methacrylate at high dose rate has disclosed

some significant departures from simple kinetic schemes of low
dose rates.

Polymerization systems which have been extensively studied
are suitable for studying the yield of propagating free radicals as a function of the absorbed dose rate.  Methyl methacrylate is such a system:  kinetic rate constants have been
extensively investigated.  Termination is bimolecular by both
disproportionation and combination mechanisms.  Since the probability of bimolecular termination is directly proportional to
the square of the propagating free radical concentration, at
low dose rates and corresponding slow rates of polymerization,
the degree of polymerization will equal the reciprocal of the
chain transfer constant.  For fast reactions, the degree of
polymerization will be determined almost entirely by the probability of bimolecular termination.  For example, the chain
transfer constant of methyl methacrylate to monomer is $10^{-5}$.
At a DP of 1000, the termination probability is $10^{-3}$.  Only one
percent of the molecules are terminated by chain transfer.

The degree of polymerization is numerically equivalent to
the reciprocal of the termination probability.  This termination probability equals the sum of the probabilities of chain
transfer, disproportionation, combination and primary radical
termination.  Probabilities of chain transfer are only functions of temperature and system composition, while probabilities of bimolecular termination are also functions of the concentrations of propagating free radicals.

Over a period of time, the rate equals the product of the
number of molecules formed and the DP; also, the number of initiating radicals produced per unit time equals the product of
the number of molecules formed per unit time and the number of
initiating radicals per molecule.

If the formation of molecules by chain transfer is negligible, the number of initiating free radicals per chain equals
the reciprocal of the sum of the fraction disproportionated and
one half of the fraction formed by combination.  In methyl methacrylate the number of initiating radicals per molecule is
expected to be 1.2.

Termination by primary radicals only, would increase the
number of radicals per molecule to 2 and decrease the number
of propagating radicals per molecule to 1.  Termination by primary radicals would increase the probability of termination and
decrease the molecular weight.  For example, if half the chains

were terminated by primary radicals, the DP would be approximately one half that expected of a polymer which terminated through propagating radical disproportionation. The relatively high molecular weight observed for the polymerization of methyl methacrylate would seem to negate this possibility.

Experimental data indicate that the effective free radical yield $\phi/Z$, where $\phi$ is the efficiency of radical formation per primary event and $Z$, is the average number of polymerization initiators per polymer chain is a strong function of dose rate. This experimental fact may have been over looked in the narrow range of dose rates examined with isotopic sources. This is shown in Figure 5 with data spanning 4 orders of magnitude in each coordinate. There is no adequate theoretical explanation for the 1/4 power exponential dependance at this time but it is a clear indication that the mechanism of chain initiation is more complex than considered historically.

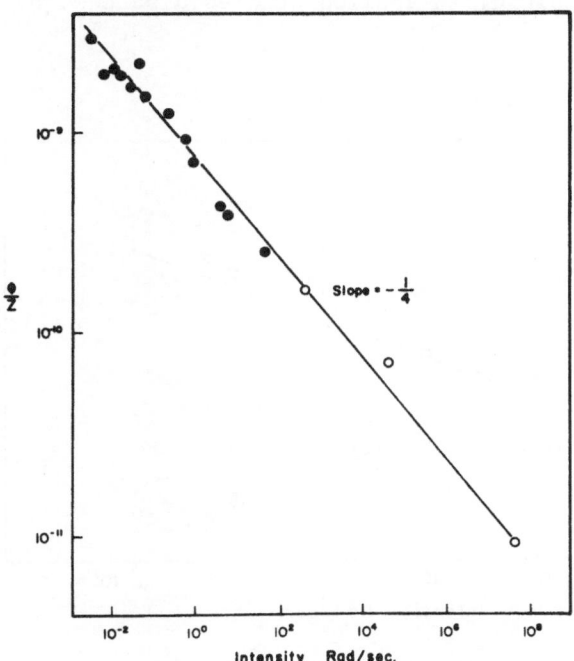

Figure 5.   Effect of Intensity on Free Radical Yield

Rederivation of the kinetics presented in Appendix 2 utilizing the empirical function yields the dual plot Figure 6 after Chapiro with modified slopes in the rate and molecular weight dependency in good agreement with the MMA data obtained.

Space does not allow the full examination of the possible impact of this and other kinetic aspects of electron beam acceleration but the following generalizations should be recognized. With scanned beams, both one dimensionally and two dimensionally scanned beams, there are significant dose rate-time effects. All approaches of dosimetry, difficult in themselves, lead to time average dose rates. The beam itself a distribution of intensities follows complex patterns. While it is true that experimental programs have found rather good correlations with time average dose rates as discussed by Hoffman[5] in his excellent fundamental review article, processes non-linear in dose rate such as described for MMA can lead to deviations from overly simplified kinetic models particularly in these non-steady state applications. Unsteady treatment is possible and leads to limiting values of the conditions, as determined by the kinetic time scale of the chemical events. This approach should to be helpful in applied studies discussed by Miranda and Huemmer[6].

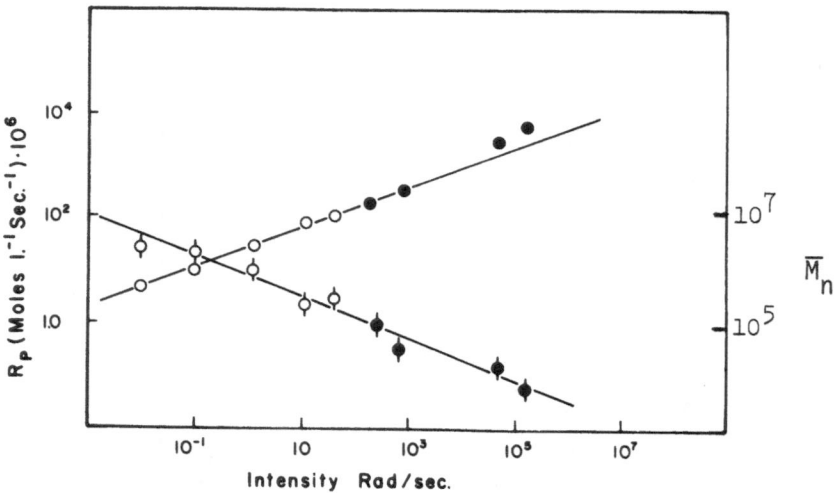

Figure 6.   Rate and Molecular Weight as a Function of Intensity

Mathematics describing various characteristics of scanned beam kinetics are developed in Appendix 3. Equations estimating post effects and scan frequency effects are presented.

## POLYMERIZATION OF METHYL METHACRYLATE WITH DI AND TRI METHACRYLATES

Methyl methacrylate copolymerizes in the presence of poly-methacrylates to form polymer chains with pendant double bonds. These unsaturated pendant units may react with propagating free radicals to form cross-links. The crosslinking reactions may form a gel or an insoluble phase as the reaction proceeds. The conversion necessary to produce gel, $\theta_c$, may be estimated by the following equation:

$$\theta_c = [\gamma \rho Y_w]^{-1}$$

where $\gamma$ is the product of the relative reactivity of the poly-functional methacrylate and the relative reactivity of the pendant group, $\rho$ is the fraction of molecules which are poly-functional, and $\bar{Y}_w$ is the weight average degree of polymerization.

After gelation the polymerization medium becomes more viscous. Free radical mobility is reduced, decreasing the rate of termination. Molecular weights, rates of polymerization, and free radical concentrations are increased through the "gel effect".

These kinetic features of the "gel effect" are highly desirable to produce non-tacky coatings at relatively low doses of radiation. Unfortunately, the conversion necessary to gel varies inversely with the molecular weight as seen in the previous equation and the molecular weights of polymer formed at high dose rates are relatively low. The concentration of poly-functional methacrylate may be increased to induce gelation at low conversions.

At a dose rate of 2.8 Mrads per sec. methyl methacrylate was 4% converted after 4.2 seconds. Under the same conditions, methyl methacrylate and ethylene glycol dimethacrylate mixtures of 77% and 53% dimethacrylate formed a hard solid. Methyl methacrylate with 32% ethylene glycol dimethacrylate was 30% converted during this time period. Figure 7 shows the relationship between the conversion of monomer to a solid and the concentration of ethylene glycol dimethacrylate at two dose rates.

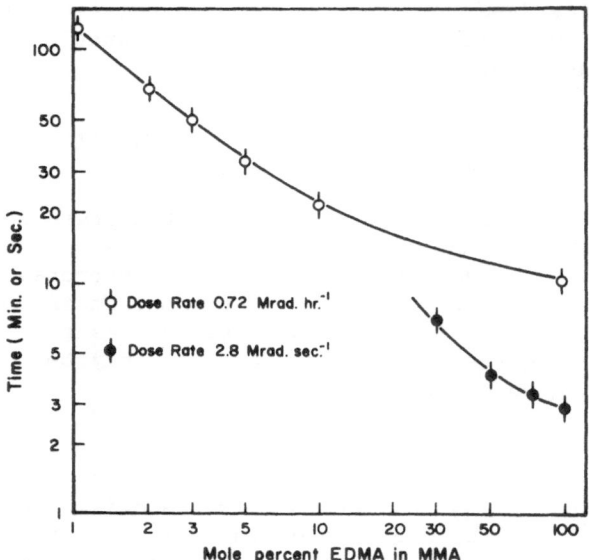

Figure 7.   Effect of Ethylene Dimethacrylate Concentration on
            Time Required for Complete Conversion.

The reaction time varied approximately with the inverse of
the square root of dose rate in concentrated solutions of
EGDMA, but relatively longer times were observed at high dose
rates for lower concentrations of EGDMA.  At high dose rates
the reduced molecular weight of the primary chain delayed the
"gel effect".

     The gel effect also has a pronounced influence on the
decay of free radicals after irradiation.  The time constant
for polymerization after irradiation may be several orders of
magnitude greater in a gelled system.

     The effects of structure play a profound effect on the
rates and consequently the economics of curing.  High relative
reactivity reduces the time to gel and increases the rate of
generation of gel.  The effectiveness of increasing the rate
of polymerization of the various dimethacrylates studied, is
Diethylene Glycol Dimethacrylate > Diisopropylene Glycol Di-
methacrylate > 1,3 Butylene Dimethacrylate > Ethylene Dimeth-
acrylate > 1,10 Decamethylene Dimethacrylate > 1,6 Hexamethy-
lene Dimethacrylate > Dimethylol Propanol Dimethacrylate > Te-
tramethylene Dimethacrylate.

The trifunctional monomer, 1,1,1 Trimethylol Propane Tri-methacrylate, was even more effective in accelerating the rate than the dimethacrylates studied, even at the same molar meth-acrylate group concentration.  At low conversions this was thought to be due to the greater pendant density on the poly-mer backbone:  A methacrylate group on a trimethacrylate which reacts will add two pendant unsaturated groups on the polymer backbone, whereas only the pendant group is added as a dimeth-acrylate unit is copolymerized.

## COMMENT

Marked departures from kinetic behavior normal at low dose rates are encountered.  These high dose rate effects are com-pounded by strong heat effects.  Careful experimental programs can define limiting behavior and model calculations based on these will enable performance to be predicted for several spe-cific cases.  Further, model calculations often enable the se-paration of these individual effects.

## REFERENCES

1 Williams, F., "Fundamental Processes in Radiation Chem-istry," Ed. P. Ausloos, Interscience Pub., N. Y., pp. 515-598 (Chapt. 8), 1968.

2 Bech, H. and Plumer, F., "Strahlungspolymerization von Akrylnitril Mit Schnellen Electronen in einer Umlaufap-paratur," Plaste und Kantschuk, 17, 2, 80 (1970).

3 Chapiro, A., "Radiation Chemistry of Polymeric Systems," Interscience Pub., New York, 1962.

4 Squire, D. R., Cleaveland, J. A., Hossain, T.M.H., Oraby, W., Stahel, E. P. and V. T. Stannett, "Studies in Radi-ation Induced Polymerization of Vinyl Monomers at High Dose Rates, Part I Styrene," J. Applied Poly. Sci., 16, 645 (1972).

5 Hoffman, A. S., "Electron Curing of Coatings:  Present Status," Proceeding IAEA Meeting, Seoul, Korea, Sept. 28, 1970.

6 Miranda, T. J., and Huemmer, T. F., "Radiation Curing of Coatings," J. of Paint Tech. 41, 529, 118 (1969).

ACKNOWLEDGEMENT

We would like to thank the Army Research Office (Durham) for their partial support of this work.

APPENDIX 1

Heat Transfer in Reacting Films

Heat transfer could be a limiting factor in the curing of films if a temperature increase occurs. The film temperature distribution during cure could modify the film properties and limit the curing rate.

A limiting case of this heat transfer problem will be presented: quasi steady state where the maximum temperature in the film is not high enough to influence the reaction rate. The energy generated in each volume element of the film may be considered constant in this limiting case. The total energy generated is the sum of the energy generated by reaction and energy added by radiation.

$$\Delta E_T = \Delta HR_P + \Delta q_{rad}.$$

The temperature distribution will then be a solution of

$$-\frac{\partial^2 T}{\partial X^2} = \psi \quad \text{where } \psi = \frac{\text{energy generated per unit volume time}}{\text{thermal conductivity}}$$

subject to the boundary conditions of the particular geometry of interest. Consider the radiation induced curing of a coating on a constant temperature $T_1$ substrate. The ambient temperature $T_A$ is also assumed constant. The heat transferred to the ambient gas may be represented by an empirical equation

$$\frac{d\Delta E_T}{dt} = K_a \,(T_A - T_2), \tag{1}$$

Where $T_2$ is the temperature at the surface of the film, $K_a$ is the heat transfer coefficient for the surface cooling, and $\Delta E_T$ is the quantity of thermal energy transferred into the film.

The thermal energy entering the surrounding gas may be equated to the thermal energy leaving the film, yielding the following relationship

$$\frac{dT}{dX} = \frac{K_a}{K} (T_a - T_2)$$  (2)

K is the thermal conductivity of the film.

Temperature distributions in the film may be calculated from solutions of the differential equation

$$-\frac{d^2T}{dX^2} = \psi,$$  (3)

where $\psi$ is the ratio of the energy generated per unit volume from reaction and the thermal conductivity.

Integration of Eq. 3 yields

$$-\frac{dT}{dX} = \psi X$$  (4)

Combining Eq. 2 and 4 at $X = X_2$, the following relationship develops.

$$-\psi X_2 = \frac{K_A}{K} (T_A - T_2)$$  (5)

Integration of Eq. 4 permits the evaluation of the distance between $T_2$ and the maximum temperature in the film.

$$X_2^2 = \frac{2}{\psi} (T_M - T_2)$$  (6)

Simultaneous solution of Eqs. 5 and 6 gives the desired result

$$X_2 = -\frac{K}{K_A} + \sqrt{\frac{K^2}{K_A^2} - \frac{2}{4} (T_A - T_M)}$$  (7)

The distance between the maximum temperature and the substrate may be obtained by integrating Eq. 4.

$$X_1 = \pm \sqrt{\frac{2}{\psi} (T_M - T_1)}$$  (8)

The total thickness is the sum of $X_1$ and $X_2$

$$X = -\frac{K}{K_A} + \sqrt{\frac{K^2}{K_A^2} - \frac{2}{\psi}(T_A - T_M)} \pm \sqrt{\frac{2}{\psi}(T_M - T_1)} \quad (9)$$

If $\psi$ is constant and the rate of heat lost to the atmosphere is small compared to the heat flux to the metal substrate, the temperature distribution in the film may be approximated by

$$T \cong T_S + \frac{1}{2}(2.39 \cdot 10^{-6} \frac{cal}{g_m \ rad}) \frac{I}{K} \rho \ X^2$$

where X is the distance from the substrate at temperature $T_S$. This approach assures that the heat generated by reaction is relatively unimportant.

For a $2.5 \cdot 10^{-2}$ cm. organic coating cured at 3 Mrad/sec., the maximum temperature obtained would be approximately 10°C greater than the substrate. The steady state maximum temperature is influenced more by the thickness than the intensity. Calculations show that the temperature at the surface will be only slightly less than the maximum temperature in the film. For thick films at high dose rates the maximum temperature may be considered approximately the adiabatic temperature (thicker than one mil at doses less than 2 Mrad).


APPENDIX 2


Dose Rate Efficiency

At high dose rates the efficiency of chain initiation decreases. The model which is proposed is

Initiation:

$$M \rightsquigarrow \dot{M} \qquad\qquad R_i = \phi_o I^A [M] \qquad\qquad (1)$$

Propagation:

$$\dot{M}_n + M \rightarrow \dot{M}_{n+1} \qquad R_p = k_p [M][\dot{M}] \qquad (2)$$

Termination:

$$\dot{M}_n + \dot{M}_M \rightarrow M_{n+m} \qquad R_t = k_t [\dot{M}]^2 \qquad (3)$$

Assuming steady state, i.e. $R_i = R_t$

$$\dot{M} = (\phi_o I^A [M])^{1/2} K_t^{-1/2} \qquad (4)$$

$$R_p = KpK_t^{-1/2} [M] \, (\phi_o I^A)^{1/2} \qquad (5)$$

$$\bar{P}_N = zKp[M]^{1/2} \, K_t^{-1/2} (\phi_o I^A)^{-1/2} \qquad (6)$$

Solving Eqs. 5 and 6 for the ratio $K_p \, K_t^{-1/2}$ and equating the results yields upon rearrangement:

$$\phi = \phi_o I^{A-1} = ZR_p \, \bar{P}_n^{-1} \, I^{-1} \, [M]^{-1} \qquad (7)$$

$$(A-1)\ln I + \ln \phi_o = \ln \phi \qquad (8)$$

If the proposed model is valid, a graph of $\ln \phi$ vs. $\ln I$ should have a slope of A-1.0 and an intercept of $\ln \phi_o$ (Fig. 4).

## APPENDIX 3

### Non-Steady State Polymerization Kinetics

The unsteady state response of a polymerizing system may be predicted using the mathematical model described in Eqs. 1 - 3. Two approaches to the derivation of the desired equations will be adopted. The first is an algebraic relation between $\phi$ and the measured system variables and the second is a relationship between the intensity, irradiation times, and expected rate of polymerization and degree of polymerization obtained through integration of the rate equations.

If N is the number of polymer molecules formed over a certain time span, then z N is the number of moles of initiated radicals. Since $[M] \, \phi I$ is the rate of initiation of radicals, $[M]\phi It$, equals the moles of initiated radicals generated during a time span $t_1$, assuming that the rate of generation of radicals during the time span $t_1$ is constant.

$$[M]\phi It = ZN \tag{9}$$

If the radiation occurs intermittently with a "light" time of $t_1$ and a "dark" time of $t_2$, and the magnitude of $t_2$ is large enough for the free radical concentration to be much less than the concentration at the end of $t_1$, the moles of monomer converted during the time span $t_1 + t_2$ is $N\bar{P}_n$ and the rate of polymerization is

$$\bar{R}_p = N\bar{P}_n (t_1 + t_2)^{-1} \tag{10}$$

Substituting for N from Equation 32 and rearranging gives

$$\phi = \frac{\bar{R}_p Z (t_1 + t_2)}{\bar{P}_n I[M] t_1} \tag{11}$$

Equation 11 would be numerically equivalent to Equation 7 if $\phi$ did not change with the radiation intensity, since $I t_1(t_1 + t_2)^{-1}$ is the experimentally measured time average intensity. The ratio of the rate to the molecular weight should be a constant for constant values of $\bar{I}$, though the rate should change if the ratio of the "light" to "dark" times or the frequency was changed.

The product of the rate and degree of polymerization should give a measure of the temperature effect expected since the "conversion mean" of $K_p^2 K^{-1}$ may be calculated from experimental variables.

$$RP \cdot DP = k_p[M][\dot{M}] \cdot \frac{2K_p[\dot{M}][M]}{K_t[\dot{M}]^2} = \frac{2K_p^2[M]^2}{K_t} \frac{[\dot{M}]^2}{[\dot{M}]^2} \tag{12}$$

The ratio of the square of the average to the average of the square may be calculated from "rotating sector method" equations.

During the "light" period $t_1$, the derivative of the free radical concentration may be written

$$\frac{\partial \dot{M}}{\partial t} = \phi I[M] - K_t[\dot{M}]^2$$

Integrating from $t = o$ to $t = t_1$

$$\frac{\dot{M}}{\dot{M}_s} = \text{Tanh} \frac{t}{\tau} \tag{13}$$

where $\tau \equiv (K_t [\dot{M}_s])^{-1}$

The ratio of the time average rate of polymerization to the steady state rate of polymerization may be written as follows:

$$\frac{R_p}{R_{p_s}} = \int_0^{t_1} \frac{\dot{M}}{\dot{M}_3} \, dt = \int_0^t \tanh \frac{t}{\tau} \, dt \qquad (14)$$

Integrating,

$$\frac{R_p}{R_{p_s}} = \frac{\tau}{t_1} \ln \cosh \frac{t_1}{\tau} \qquad (15)$$

for values of $t_1$ where $t_1 < 10\tau$ the corresponding value of $\cosh \frac{t_1}{\tau}$ may be found tabulated in various references. For values of $t_1$ greater than $10\tau$ the relationship may be simplified.

$$\frac{R_p}{R_{p_s}} = 1. - 0.68 \, \tau t_1^{-1} \qquad (16)$$

During the "dark" period the rate of radical decay may be written

$$\frac{d[\dot{M}]}{dt} = -K_t [\dot{M}]^2$$

Integrating from $t = t_1$ to $t = t_2 + t_1$,

$$\frac{[\dot{M}]}{[\dot{M}]_{t_1}} = [1 + \tau^{*-1} t_2]^{-1} \qquad (17)$$

where $\tau^* \equiv (K t [\dot{M}]_{t_1})^{-1}$

It is interesting to note that the time constant for the decay of radicals, $\tau^*$, is increased if the concentration of radicals at $t_1$ is less than the steady state concentration. This means that the "post effect" would have more relative significance in the overall polymerization rate as $(t_1 + t_2)$ is decreased.

Since

$$\frac{R_{pt_2}}{R_{pt_1}} = \int_{t_1}^{t_2 + t_1} \frac{\dot{M}}{\dot{M}_{t_1}} \, dt \tag{18}$$

$$\frac{R_{pt_2}}{R_{pt_1}} = \frac{\tau^*}{t_2} \ln \left(1 + \frac{t_2}{\tau^*}\right) \tag{19}$$

$$\tau^* = \tau \left(\text{Tanh} \frac{t_1}{\tau}\right)^{-1} \tag{20}$$

These equations are valid for intensity frequencies low enough to assume that $[\dot{M}]_{t_2} \cong 0$.

# THE PRESENT STATUS OF ELECTRON BEAM CURING OF COATINGS

Kennard H. Morganstern, Ph. D.

Radiation Dynamics, Inc.

1800 Shames Drive, Westbury, N. Y.

## INTRODUCTION

Electron beam curing of coatings is now approximately a decade old and as a technique is demonstrating the same lengthy gestation period that has been evident with respect to other radiation process techniques.

In the latter instance, in spite of twenty years of development work and substantial quantities of money spent, only now are meaningful industrial process applications coming to the forefront. Significantly, the acceptance of radiation as a new process technique is due, in the main, to the recognition on the part of industry that radiation is a practical process with demonstrable economic advantages. Unfortunately, during its early growth, most people were looking for unique and startling advantages and it was only after the recognition that radiation can do the same job but perhaps faster and cheaper that acceptance began to take place.

It appears that this same situation exists in the field of electron beam curing of coatings. Rather than looking for the extraordinary, it appears that one should be satisfied in being able to produce equivalent coatings but with economic advantages.

## BACKGROUND

RDI initially got involved in electron beam curing of coatings with a contract in 1962 from U.S. Plywood. U.S. Plywood's interest was in the possible application of electron beam for the curing of clear coatings on wood panels. At that point, we had installed in our radiation service facility a 1.5 MeV, 10mA Dynamitron which was used as the radiation source for this work. Most of the coating systems investigated were clear glossy coatings and we were able to demonstrate excellent cures but only in the absence of oxygen. The elimination of oxygen was accomplished by product irradiation in an inert atmosphere and through the use of release sheets. Since U.S. Plywood felt that inert atmosphere or release sheets would present production problems, a substantial portion of time was spent in trying to develop a coating system which would radiation cure under ambient conditions. Unfortunately, we never succeeded in getting the equivalent cure in the presence of oxygen, and after a year the program was terminated. The conclusion reached was that this type of work was better suited to experimental groups with a stronger competence in the radiation chemistry area.

About the same time, Dr. William Burlant, working at Ford Motor Company, approached the problem from the standpoint of the basic interaction of radiation of various chemical systems, and his early work finally resulted in a number of patentable systems.

In 1965, RDI became convinced that radiation curing of coatings was a viable process technique; naively it was felt that all that was lacking at that point was acceptable coatings. Consequently, a low voltage-high current accelerator was introduced, designed specifically for electron beam curing of coatings. This unit was designated as the Dynacote, and the first such device of 300 kV, 25mA was installed in RDI's radiation service facility at the end of that year.

Substantial amounts of publicity were given to the

introduction of the Dynacote, with the net result the
company literally got swamped by potential end product
users, who served to reinforce the generally accepted
feeling that radiation curing of coating made sense.  In
addition, a number of chemical and paint companies
instituted development programs using the Dynacote in
our radiation service facility.

It soon became evident in observing the development
activities going on in the service facilities that although
RDI had brought systems 90% of the way in 1962, the last
10% was not going to be easy to achieve.  And it was
obvious that electron beam curing of coatings would not
become an acceptable procedure until suitable coating
systems were developed and made commercially available for a variety of substrates.

In November of 1966, Ford Motor Company licensed
Boise-Cascade under their Electrocure System, and
installed equipment in Boise, Idaho for pilot plant production of radiation coated wood panels.  This first semi-
commercial facility aroused wide interest and appeared at
that point in time to be the precursor to rapid acceleration
in the field.  Unfortunately, however, the process at Boise-
Cascade did not offer the advantages originally anticipated and because of ample capacity on more conventional
curing lines, Boise-Cascade shut down the facility.

About the same time the Boise facility was going
down, Ford set up the first true radiation curing production facility at their Saline, Michigan plant.  This facility, employing a multiplicity of 300kV electron beam
generators, was designed to handle the curing of coatings
on the dashboards of Ford Motor Company cars.  The
facility has worked extremely well since its activation
in 1970, and has a capacity of curing as many as 30,000
units a day.  The substrate is an ABS and the application
is a natural since conventional heat curing of this type
of substrate would by necessity be slow.

The expansion of the Saline facility this year with

additional beam capacity testifies to the success of
this first commercial venture. In November of 1970,
Ford also installed an Electrocure System on a coil
line for Wolverine Pentronix. The facility at Wolverine
is designed to handle coil coatings at a few hundred
feet per minute and is in line with a conventional heat
curing system. Therefore, either one or both of
these mechanisms can be employed.

It is obvious from this historical review that Ford
has been very active in the field both from the standpoint of chemical systems as well as radiation generating
equipment.

However, many of the other major paint suppliers
also have had active radiation programs throughout
this period. PPG through its subsidiary Radiation
Polymer, Inc., has been actively pursuing the development of coating systems for wood, metal and plastic
substrates, and they claim now to have commercially
available systems for wood, and to be very close for
other substrates. O'Brien Paint Company, through
its subsidiary Beta-Cure, has likewise developed substantial technical know-how in the coatings field and is able
to offer commercially available systems for a wide
variety of substrates. In addition, Celanese Corporation,
Glidden Paint, and Sherwin-Williams have also been
active in investigating coating systems.

Obviously, the potential advantages of radiation curing
were not lost to our friends overseas. In Europe a
number of major lacquer firms have been active in the
field and at the same time Ford has established a
licensee in Holland, Sikkens, a member of the Akzo
group. In addition, Ford is about ready to announce the
opening of the first commercial radiation facility in
Europe. This facility will be used for various coatings
on a variety of wood products. In Japan, the Japanese
Atomic Energy Research Institute at both Osaka and
Takasaki have developed paint curing programs in this
area;   a joint venture between Kansai Paint and

C. Itoh, called Nippon Electrocure, was formed as
a Ford licensee.  PPG and Mitsubishi have also
established a joint venture in Japan based upon PPG's
developmental efforts here in this country.  Further,
Toray Ind. has been actively investigating radiation
curing applications.

This listing of companies actively investigating
the radiation curing of coatings is not meant to be all-
inclusive, but is indicative of the broad base and
continuing activity going on world-wide, and should
auger well for the future.

## ADVANTAGES

The continuing interest in radiation curing has re-
mained, in spite of the large amount of effort required,
because there appear to be definite advantages that
would accrue to the user of this process technique as
proper coating systems become available.

The first advantage would be the rapid speed with
which one could cure coatings.  With conventional heat
curing systems, one is naturally  limited by practical
considerations such as oven temperatures, oven length
and problems of supporting the product through the oven.
Because radiation curing is, for all intents and purposes,
instantaneous, one can process products at speeds which
are limited only by the requirements imposed due to
procedures before the curing element in the system--such
as painting, cleaning, etc.  Certainly in applications such
as coil coatings one can envisage speeds of 500 to 1000
feet a minute.  These speeds entail substantial milli-
amperes of electron beam current, but this is readily
achievable and more practically achievable than the oven
lengths that would be required for the equivalent speeds.

The second advantage relates to costs.  Generally,
one finds that radiation curing is less expensive than an
equivalent heat curing system. Part of the economic
advantage arises due to the fact that one can, with modest

cost additions, add to the electron beam power of the
system, thereby getting proportionately higher production
through-put with a minimal amount of additional capital
investment.

Adding to the cost advantage is the fact that radiation
curable systems are generally solvent-free, and as a
result the costs of the solvents, which have to be paid for
in conventional coating systems, even though they go up the
stack, are avoided in a radiation curable system.

And finally, with radiation one can achieve a much
higher degree of energy utilization than is possible with
conventional heat curing, thereby bringing about a reduction in operational costs.

Third, it is obvious from what was indicated earlier,
that a radiation facility requires less space for an equivalent through-put than does a conventional system. As the
speed goes up, the disproportionality spacewise becomes
even greater. For example, in a coil coating line operating at 200 feet a minute, the oven length would be approximately 150 feet, whereas the radiation facility can be
accommodated in approximately 10 feet of line length.
For 400 feet per minute, the oven length would generally
double, but the radiation space would remain constant.

Fourth, in todays ecology conscious climate, one cannot overlook the major advantage that the electron beam
curing system provides due to the absence of solvents. In
conventional paint curing systems, as much as 60% of the
total can be solvents which are released into the air during
the drying process. In order to comply with the Environmental Protection Agency's new air quality standards, costly
anti-pollution equipment has to be installed on the drying
ovens. Although there will be some volatiles given off with
an electron curable system, the quantities are substantially less than with a conventional system, and in all likelihood will comply with the Federal Air Quality Act as it
now stands without any stack requirements. Although with
radiation there is the potential problem of ozone genera-

tion and removal. This can be handled rather inexpen-
sively by heat systems that recondition the ozone with
oxygen or by chemical solutions which absorb the ozone.

There are still other advantages which one can
envisage related to electron beam curing facilities. However, the four above represent the main ones and certainly are important enough to warrant the extensive
efforts that have gone on and are still going on to bring
this process to full commercial fruition.

## DISADVANTAGES

Obviously, one cannot have a new process that is a
winner all the way. There are some problems attendent
to the electron beam curing technique which have to be
considered.

First, the capital cost involved in both radiation
equipment and its radiation environmental shield is
somewhat higher than costs customarily attributable to
the curing element in a paint line. However, as these
electron beam devices continue to get more powerful,
thereby having more product throughput, the disparity
between the electron beam equipment costs and an oven
of equivalent capacity begins to evaporate.

Second, perhaps the single largest disadvantage is
the requirement for new paint systems which are radiation
curable. As indicated earlier, obviously this is not
easily accomplished and limitations are imposed due to the
mechanism of radiation curing. However, the fact that
there are commercial systems available and newer ones
coming on stream shows that the technology is moving
ahead.

Perhaps the single biggest obstacle to electron beam
curing is the natural reluctance on the part of the end user
to try something new - rather than something that is
already accepted. However, as more facilities are estab-

lished, there should be a proportionate reduction in this
natural hesitancy.


## TECHNICAL ASPECTS

There is a basic relationship between radiation dose
necessary to cure a coating, the milliamperage of the
equipment, and the process speed. This relationship
can be expressed as:

1 mA = 1000 square feet per hour per megarad

A megarad is the unit of dose generally used in this
field and is equal to a million Rads: a Rad is equal to
100 Ergs per gram.

The energy of the electron is dependent upon the
thickness of the coating which one wants to cure. The
thicker the coating, the more energetic the electron
must be. However, for coating thicknesses up to 10
mils, 300 KeV generally is sufficient. For thinner coatings on the order of 1 to a few mils, 200 to 250 KeV
would be still more acceptable. Although one would like
to use as low a voltage as possible based upon the coating thickness, there are practical limitations imposed
on the choice of voltage. These limitations are primarily
due to problems of bringing the electron beam from the
evacuated portion of the accelerator onto the product.
The interface between the two is generally a thin metallic
window. This window must maintain vacuum integrity and
the pressure differential and have sufficient thickness for
mechanical strength at elevated temperatures, while
being sufficiently thin to minimize absorbed beam energy.
In addition, the air column between the window and the
product adds to the required voltage.

Generally, one first defines the practical voltage
necessary to accomplish the curing desired, the milliamperage is then determined by the product speed and
the dose requirement based upon the equation indicated

above. How this relationship relates to the economics of electron beam curing will be discussed later.

A number of types of low voltage - high current equipments have been developed and are available from a variety of sources. Generally these products are 200 to 300 kV with milliamperage as high as 200. This equipment includes RDI's Dynacote, the ICT, the Tiger, Ford's equipment, and a variety of similar devices now becoming available in both Europe and Japan.

Essentially, all of these systems operate on the same principle. A high voltage power supply is cable connected to the acceleration column. The acceleration column consists of an electron gun which emits a stream of electrons and an evacuated acceleration tube down which the electron particles are accelerated. Generally, the electrons are focused as a pencil beam and magnetically scanned to emerge as a curtain of electrons through a thin window. In one instance, however, rather than using a magnetically scanned electron beam, the equipment employs a linear filament which provides the electron curtain. In each instance, because of the curtain arrangement, the electron beam equipment is most ideally suited for relatively flat material.

In the future, because of the requirements for higher beam current in order to operate at higher speeds, and varying product configurations, one can envisage electron beam configurations similar to those that exist in infrared ovens. RDI has been working on the development of a new product which we call the "XBD", Expanded Beam Dynacote - which is designed as the next generation of electron beam coating equipment.

## ECONOMICS

As indicated earlier, one of the major reasons why electron beam curing is still of interest is the potential economic advantages that accrue to the user of this system.

Obviously, the economics improve as the product through-
put goes up.  Since product through-put is related to dose
requirements, this imposes the requirement on the paint
system to be cured at relatively low dose levels.  If, for
example, one has a system that will cure at 2 megarads,
and based upon the operational cost per hour of some of
the newer high powered electron beam systems, one can
project curing cost on the order of 10 to 20 cents per
thousand square feet.  I believe if one compares this
with conventional heat curing systems, one will find that
this is a substantial reduction in curing costs.

## SUMMARY

Although the electron beam curing of coatings is no
longer in its infancy, and although most of the initial
enthusiasm which greeted the early announcements has
diminished with time, nevertheless, for the reasons
cited above, it is evident that radiation curing will take
its place along side of other techniques as a viable curing
system.

# ELECTRON BEAM PROCESSOR TECHNOLOGY

S.V. Nablo, J.R. Uglum and B.S. Quintal

Energy Sciences Inc.
111 Terrace Hall Avenue
Burlington, Massachusetts 01803

## INTRODUCTION

The generation of powerful electron beams for industrial process application has evolved from the "point source" or cylindrical beam geometry required for early radiation research and radiographic applications. "Powerful" in this sense would refer to power levels above a few kilowatts at energies of 100 keV or greater. Electron generation and acceleration are accomplished in an evacuated tube along which the accelerating voltage is distributed by means of a resistive divider; this is typically located in the pressurized environment surrounding the accelerator tube. Gases such as $SF_6$ or $CO_2 + N_2$ are used to insulate the voltage generator and accelerator structure. This cylindrically symmetric tube structure is used to deliver a pencil beam to the exit window or metal foil through which it emerges into the ambient environment, where it can be usefully applied to the workpiece. For strip application where uniform illumination of a two dimensional surface is required, an electrostatic or magnetic scanner is added which serves to distribute the cylindrical beam over a one or, if necessary, two dimensional scan. Such a configuration is shown schematically in Figure 1 (a) and is referred to as a hybrid (vacuum/pressure) approach to accelerator design.

Although this has been the conventional route to energetic industrial processor development, it has been more the result of the initial emergence of pencil beam machinery in the 1940's which was then adapted to strip application, than the result of directed design. Several disadvantages of this approach are immediately obvious. Perhaps the most severe is the very large surface area which must be shielded to permit safe operation of the processor. Any surface enclosing the evacuated acceleration-scanner volume can serve as a source of Bremsstrahlung or X-rays generated by scattered electrons which are continuously created along the entire length of the system. As a consequence, one of the first considerations of efficient processor design must be that of shielding volume, hence the need to minimize machine surface area. The configuration of 1 (a) results in a large evacuated volume with its associated high pumping capacity requirements and size/weight penalties.

Of greater significance to the radiation chemist may be the very high instantaneous dose rates experienced within the relatively small area illuminated by the beam as it is scanned over the film or coating surface at audio-frequencies. For example, a 1 cm diameter beam scanned in one dimension over a 1 meter window will provide a dose rate two orders of magnitude higher than an unscanned strip system, working at the same average power. For a typical chain polymerization curing application the results of this elevated energy delivery rate can be quite profound: the rate of cure or polymerization is proportional to the square root of dose rate where free radical recombination occurs by bimolecular termination. As a consequence, the dose to cure (in effect, the energy investment required to complete the process) will increase roughly as the square root of the instantaneous dose rate at which the resin/monomer system receives energy.[1] The implications to the economics of

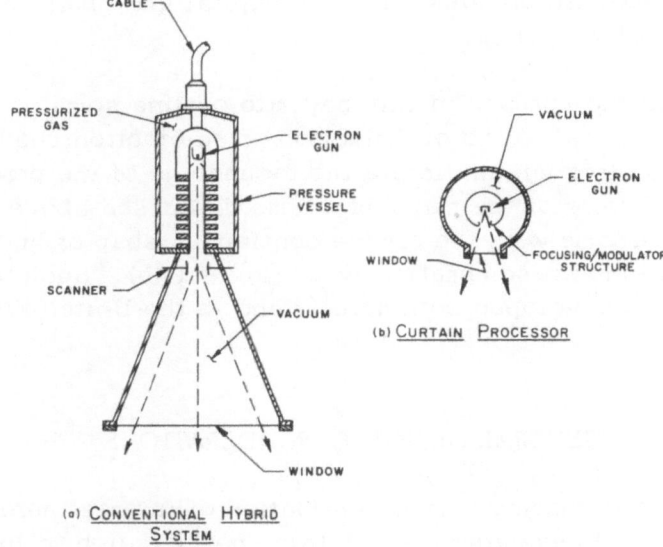

Figure 1

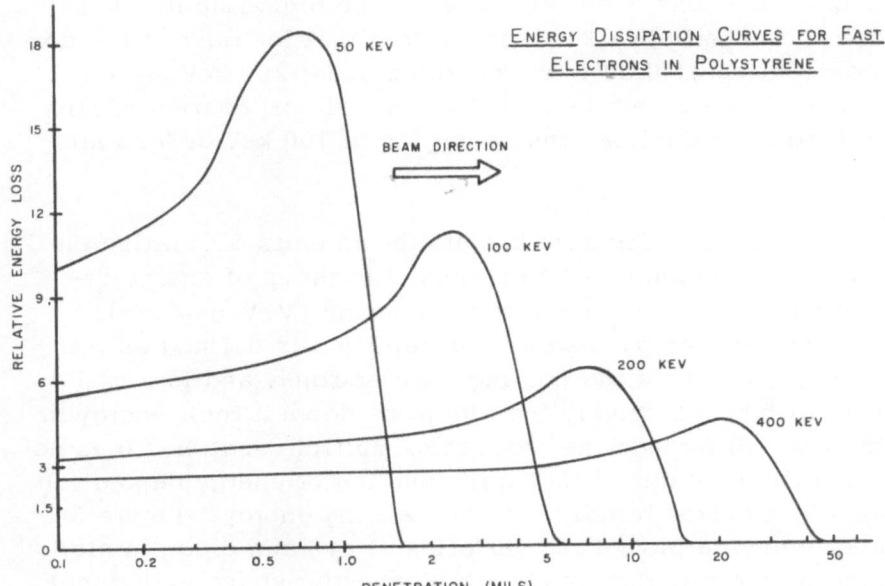

Figure 2

the process are obvious - both in capital and operating cost
of the curing system.

It is the purpose of this paper to outline some of these
factors critical to the optimization of an electron beam process - factors which dictate the technology of the processor
design. Some of the major problems discussed above have
been overcome with the simple continuous strip or curtain
geometry shown schematically in Figure 1 (b). Such units
have been developed both here[2] and in the United Kingdom.[3]

## GENERAL DESIGN CONSIDERATIONS

For the coating curing application of interest here, it is
instructive to examine the electron energy region of interest.
In a cursory manner, we will examine the processor requirements working back from the coating. Some typical range-
energy relations[4] for electrons in the region up to 400 keV
are shown in Figure 2. It is evident that for most of the common coating applications, energies under 200 keV are entirely adequate, while for thinner metal coil coating or film
crosslinking applications, energies of 100 keV or less are
satisfactory.

The data of Figure 2 are plotted in units of relative energy loss, which may be expressed in terms of energy deposited per unit thickness of the coating ($MeV/gm/cm^2$).
This rate of energy loss with penetration is defined as the
stopping power of the coating for electrons, and is a measure of how effectively the electrons deposit their energy in
the coating through the process of multiple scatter. It is not
generally recognized that this coupling or energy deposition
efficiency rises rapidly with decreasing energy. Figure 3
shows such a plot of the variation of average electron stopping power with energy for a typical polymer (polyethylene).
It is clear that for a coating under 4 mils thickness, a 100

keV electron is more than twice as effective as a 400 keV electron and, if we exclude other losses, we can provide the same current output at only 1/4 the electrical energy expenditure, for a net energy advantage of eight.

We must examine these rather academic physical arguments with the practical considerations of getting the energy from the accelerator tube itself into the coating. Clearly, if the electron coupling efficiency in the coating is higher at reduced energies, so too will the energy losses suffered in transit increase. Figures 4, 5 and 6 attempt to summarize some of these aspects of processor optimization.

In Figure 4, the effective ranges of electrons in air at NTP are shown over the same energy range (0-400 keV) in curve A. In curve B, the path length is shown for which 50% of the beam energy is lost in the air gap between the foil window and the coating. Curves C and D present similar data for 20% and 10% loss respectively, representing acceptable conditions for a working processor. Returning to the case of our 100 keV beam, we see that air gaps of only a few cms are tolerable if these modest losses are to be realized.

Figures 5 and 6 treat the problems implicit in window design. The initially unidirectional beam from the accelerator tube will be scattered into a distribution which is a function of incident energy and foil composition as it traverses the metal window into air. For conduction cooled windows, 25 micron aluminum foil is typically used; for the simpler convection or radiativity cooled windows, 13 $\mu$ titanium foil is standard. The techniques of support and cooling will depend upon the window area and desired current density for the process of interest. These latter figures will range from a few tenths to two mA/cm$^2$ for these lower energy high rate electron processors.

From Figure 5, energy and charge loss in the window have been calculated for these two cases of interest. For

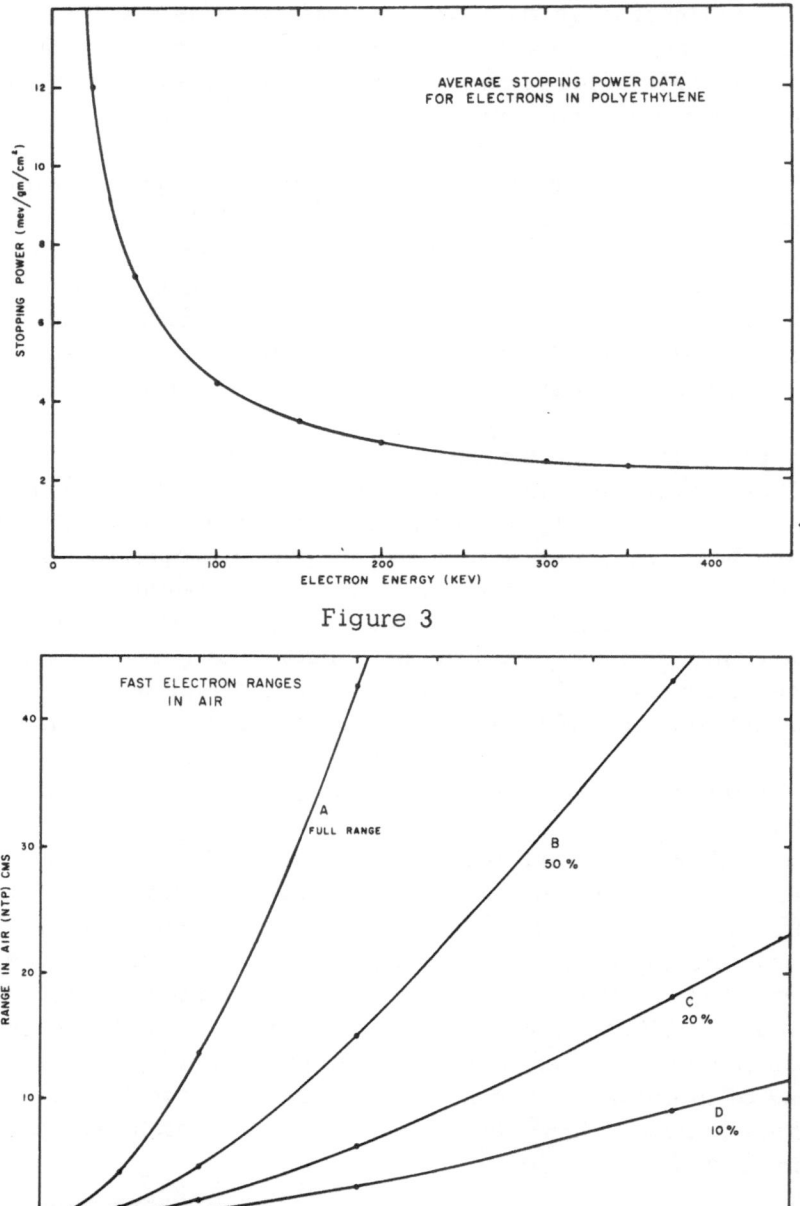

Figure 3

Figure 4

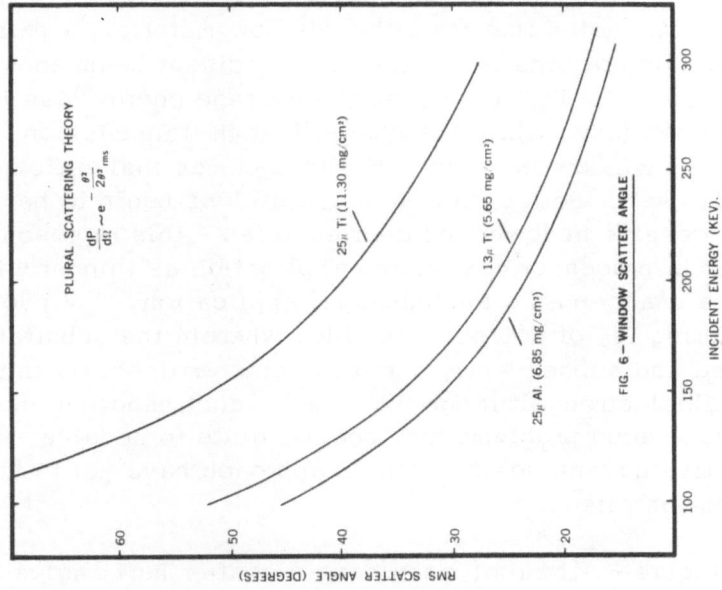

Figure 6

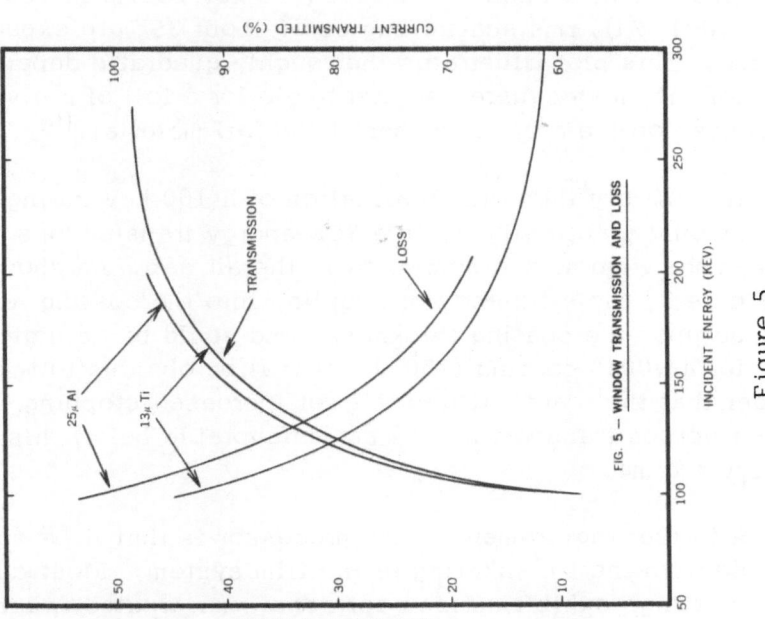

Figure 5

example, one notes that for either window material, a point of diminishing returns is realized with incident beam energies of 150 keV. For this case, the average energy loss is about 30 keV (20%) while the charge lost in transmission through the window is about 10% and becomes insignificant above 250 keV. Both current and energy loss begin to become excessive at lower incident energies - this represents a barrier in processor design and application as thinner windows are unacceptable for industrial application. "Windowless" curing is, of course, possible, wherein the substrate is coated and subsequently cured in vacuo with beams in the 30-50 kiloelectronvolt range.[5] The product handling and coating in vacuo problems now become quite formidable, and the relative advantages of such an approach have yet to be fully demonstrated.

In Figure 6, the root mean square scatter half angles for the beam as it emerges from the foil window are plotted. One sees that for the cases of interest (150 keV beams in .0005" Ti or .001" Al), rms scatter angles of about 35° are experienced. This plot illustrates the roughly quadratic dependence of the mean squared scatter angle for a foil of a given thickness on the atomic number of the foil material.[6]

In summary then, the realization of a 100 keV curing beam would practically involve 30% energy transfer loss - 20% in the window and some 10% in the air gap. As shown in Figure 2, the efficiency of coupling into the coating will depend upon the coating thickness, and could be as high as 85% for a .004" coating (100 $\mu$). It will be obvious to the reader that the practical tradeoffs of increased stopping power versus transport losses are unfavorable below this energy regime.

A further requirement of the processor is that it be compatible with inert blanketing of the film system. Most curable coatings exhibit surface softness when treated in air. This occurs because of the competition which exists between

free radical reactions with atmospheric oxygen and with unsaturated carbon-carbon bonds. Since the former terminates
chain growth, the surface region where oxygen diffusion has
occurred experiences relatively little crosslinking and remains as a low molecular weight (uncured) polymer. These
effects and their dependence upon dose-rate have been
treated at length in the literature.[7] Various techniques
are used to overcome this problem which strongly influences
the mar resistance of the cured coating. The most common
and convenient is the reduction of the oxygen content in the
curing zone to under 1% with the use of $N_2$ or hydrocarbon
blankets. For this reason, it is preferable to avoid forced
convective cooling of the window assembly which would result in unacceptable gas costs, if the inert gas was employed for both purposes. Double window techniques have
been used to separate the cooling:inert stream flow channels,
for example by the UKAEA group. Since this can involve unacceptable losses with the low energy processors of interest
for this application, conductivity (water) cooled aluminum
windows are normally used in large area industrial curing
systems. These can range from the solid cross-member grids
employed by Ford Electrocure, in which heat is removed by
conduction to the periphery of the assembly, to the more
efficient but more complex water cooled capillary structure
used on the Macpherson TIGER installation and on the Energy
Sciences Electrocurtain™ processors.

## EXAMPLES OF INTEGRATED PROCESSOR DESIGN

Bailey and Wright, in their comprehensive review of
electron beam curing,[1] have enumerated the major problems
in matching the processor characteristics to the chemical
kinetics of the coating system. Dose-rate effects and their
dependence upon the nature and concentration of prepolymer
and monomer unsaturation must be understood for each system; the dependence of dose to cure on: resin molecular
weight, the monomer properties and catalyst content must be

known and will, in turn, be dependent upon the application method and hence the coating viscosity chosen. It is of paramount importance that the surface of the cured coating be sufficiently hard to permit handling or substrate contact immediately after curing, particularly for high speed coil applications. Efforts to achieve a good compromise in the often conflicting requirements of the system have met with good success in the past five years, particularly with the polyester and acrylic resins. Nevertheless, much development work remains for the perfection of efficient coating systems to cover the multitude of applications where the speed, economy and pollution-reducing properties of the process can be used to advantage.

Figure 7 shows two views of an integrated (self-shielded) processor developed for laboratory application where simplicity of operation, and speed of sample treatment and post-cure examination are of primary importance. The system utilizes the curtain beam concept of Figure 1 (b) and is capable of operation to 200 kilovolts (with a beam energy of 175 keV at the workpiece). In this geometry, modern cathode-gun design is employed to provide ribbon type electron beams up to a meter in length with self-shielded gun diameters of under 25 cm. The system shown provides a 3 cm wide x 15 cm long beam at a dose rate of $2.10^8$ rads/second with a beam power of only 2 kilowatts. At this level of performance, it will handle coatings with a dose to cure of $2 \times 10^6$ rads at a speed of 300 cm/second (600'/minute) or a throughput of 300 ft$^2$/minute. The equivalent thermal investment required to handle this curing burden by conventional (oven) techniques would be typically 2 megawatts. If we assumed a dose to cure of 1 joule/cm$^2$ for an equivalent ultraviolet curable coating, the lamp power consumption would be over 70 kilowatts for the same capacity. The outstanding energy advantage of these electron curing systems will receive increasing attention as energy costs continue to demand a greater fraction of the finished product cost.

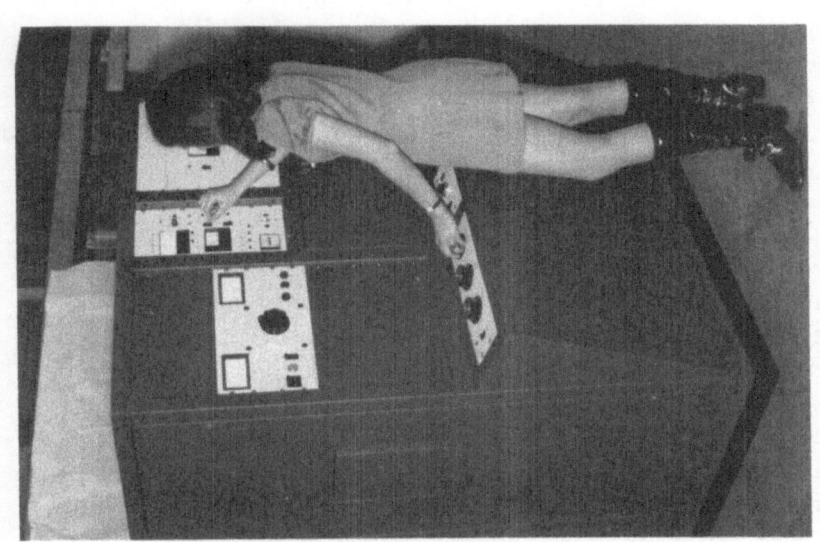

Figure 7(a) CB 150 Laboratory Unit    Figure 7(b) Cabinet Removed–Internal Configuration

The shielding features of the system shown permits continuous occupancy of the room in which the unit is operated so that it may be installed in an "open" area with only the routine use of personnel film badges for those directly involved in machine use. For the laboratory unit shown, surface dose levels are below 0.5 mr/hour at full power with the use of 6 mm lead shielding, some of which is shown in Figure 7 (b) around the sample tray cassettes and conveyor tunnel. The compact design of the curtain type processor for coatings/film applications permits efficient "shielding at the source" rather than the conventional volume or vault approach traditionally accepted as a necessary penalty in using this form of energy. For the 2 kilowatt processor, for example, full shielding is accomplished with less than 200 kilograms of lead. Comparable savings have been demonstrated on the larger units at elevated power levels.

These units, although designed for continuous application, can be operated in a pulsed mode for use in reaction kinetics studies; e.g., in simulating those dose rates characteristic of the scanned beam machines depicted in Figure 1 (a). For example, the unit of Figure 7 can be operated at one ampere output with a pulse width of 100 $\mu$secs to simulate the $2.10^{10}$ rads/second instantaneous rates of a scanned unit.

A higher power scanned unit of integrated design is shown in Figure 8.[8] Called the "Electron" by its manufacturers,* it provides a scanned (50 Hz) beam of up to 1.1 meters length x 4 cm in width for film/coating process application. The system provides good dynamic range with a beam energy variable from 300-700 keV with under ± 5% variation and an average beam power of 7 kilowatts. In this case the oil insulated power supply is of the cascade multiplier type and the biological shield is so constructed that

---

*The Electrophysical Institute, Leningrad, USSR.

Figure 8   700 kilovolt 7 kilowatt
self-shielded processor:
"Electron"

full area access to the processor is possible via the fully
hinged doors shown in the figure.

## CONCLUSIONS

New types of low energy electron beam processors have
been developed for the coating curing and film crosslinking
application over the past five years.  Their ease of shielding,
convenient size and resulting economies of operation have
stimulated the development of curing systems based upon
this form of industrial energy.  The flexibility of these electron beam systems now permits their use for topologically
difficult applications such as coated deep mouldings, since
beam shaping and precise control of the deposition distribution of this directed energy is now practicable.

## ACKNOWLEDGEMENTS

We appreciate the permission granted by Dr. V.A.
Glukhikh for the use of Figure 8, and the hospitality shown
to one of us (S.V. Nablo) while visiting his institute.

## REFERENCES

(1) Bailey, D.R. and Wright, A., "Electron Beam Curing", Paint Technology; p. 12, Sept. 1971; p. 23, Oct. 1971; p. 11, Nov. 1971; and p. 23, Dec. 1971.

(2) Nablo, S.V., "The Status of Electron Beam Curing (1971)" Proc. Nat. Coil Coaters Assoc. Mtg., Las Vegas, Nev., May 1971; NCCA, 1900 Arch St., Phila., Pa.

(3) Davison, W.H.T., "Process Characteristics of Electron Beam Curing", Journal of the Oil and Colour Chemists' Association 52, 946, 1969.

(4) Spencer, L.V., "Energy Dissipation by Fast Electrons", National Bureau of Standards Monograph 1, Sept. 1959.

(5) Williams, T., British Steel Corporation, Swansea, U.K., Private Communication, May 1971.

(6) Birkhoff, R.D., "The Passage of Fast Electrons Through Matter", Section 26, Handbuch Der Physik, Vol. 34, 53-162, Springer-Verlag, Berlin, 1958.

(7) Hoffman, A.S., et al., "Electron Radiation Curing of Styrene/Polyester Mixtures: II Effect of Backbone Reactivity and Dose Rate", ACS 157th National Mtg., Minneapolis, Minn., April 13-19, 1969.

(8) Glukhikh, V.A., "The Application of Charged Particle Accelerators in Industry and Medicine", Preprint P-0165, Institute for Electrophysical Apparatus, Leningrad, 1972.

# ELECTRON BEAM CURING - A NON-POLLUTING SYSTEM

Carl R. Hoffman

High Voltage Engineering Corporation
South Bedford Street
Burlington, Massachusetts 01803

The development of radiation processing systems had its inception in the late 1950's when research accelerators were employed to evaluate radiation effects on numerous materials.

At this time, the improvement in the physical properties of polyethylene was observed when exposed to ionizing radiation. Today, modification of plastic has been the principal use of radiation with over 600 Kwatts of installed accelerator power. For the future, other areas loom as potential markets for radiation processing.

In the coating field, researchers were also able to utilize this energy source for polymerization; however, radiation was not competitive with conventional methods, since both the equipment and coatings were far from production quality.

Over the years, numerous technological developments have occurred, resulting in the availability of both coating systems and equipment to process coating requirements equal to the present product quality and with numerous other advantages.

With larger more reliable systems available, radiation processes are now in use or being evaluated for use, for which earlier units were unable to process economically. Operating costs have gone down, reliability has gone up, leading to the fact that accelerator systems are being

installed at the rate of 20 per year. This is almost a
five-fold increase from five years ago.

Today, industry is confronted not only with the need to
manufacture superior quality products competitively, but
must also consider the effects on the environment.

One of the prime advantages of electron beam curing is
that coating systems can be applied and cured with almost no
solvent loss. With the introduction of stringent pollution
exhaust emission laws, radiation curable paints and coatings
will be coming into more widespread usage because of their
low stack emissions. Another point in favor of radiation
curing is the modest power input requirements compared to
the conventional heat curing ovens. Some areas of the
country already have serious natural gas shortages, and the
large volumes of gas required to heat the ovens are further
aggravating the situation. In certain instances, during
cold spells, these ovens have been forced to shut down.

Table 1 lists the many advantages for using radiation,
as well as those disadvantages that would affect overall
performance. In general, pigmentation of the coating, as
well as the type of substrate, will have only a slight
effect on the curing kinetics.

Products which would be best suited for electron beam
curing are: both indoor and outdoor wood paneling, coil
coating for cans, residential siding, mobile homes, appliances and adhesives.

The terminal voltage of the machine dictates the penetration of the  electron through the coating and into the
substrate. Since most curing operations are limited to
coating thicknesses of several mils, the machine's voltages
generally range from 225 kV to 300 kV. The exact voltage to
be used will depend on specific gravity of the coating, type
of substrate, dose uniformity required within the coating,
window design, and geometric shape of the product. Figure 1
shows dose depth curves over this voltage range.

The production throughput, for a machine, is directly
related to the beam current output, and the amount of dose
required for cure. Dose being defined in terms of energy
absorbed, 100 ergs per gram or Mrads.

Table 1.   Advantages and Disadvantages
of Electron Beam Curing.

ADVANTAGES:

   1.   Instantaneous curing mechanism.

   2.   Room temperature cure - no heat damage to
organic substrate.

   3.   Higher line speeds.

   4.   Smaller space requirements.

   5.   Better adhesion to organic substrates.

   6.   Instant start-up.

   7.   On-Off power source.

   8.   Low maintenance downtime.

   9.   More efficient use of power.

  10.   One part systems - no catalyst required.

DISADVANTAGES:

   1.   New coating formulations are required.

   2.   May require inert atmosphere.

   3.   Flat substrates are best suited.

   4.   Psychological opposition.

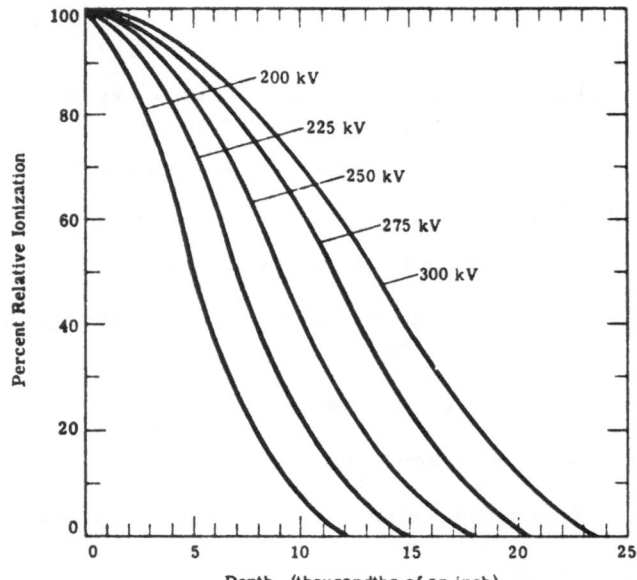

Depth. (thousandths of an inch)
Specific Gravity of Absorbing Material is = 1

Figure 1.   Depth-Dose Ionization Curves at Various
Terminal Voltages Using Standard 1 MIL
Window with 4-Inch Air Path.


The dose rate is simply a measure of the rate at which
the energy is delivered to the product.  At a given voltage,
the dose rate is a function of the beam current, irradiation
area, and an energy constant.

Thus

$$D_r = \frac{K\,I}{X \cdot Y} \tag{1}$$

$D_r$  =  dose rate, Mrads per minute
$I$  =  beam current output, mA
$Y$  =  scanning width, meter or feet
$X$  =  beam width at product, meter or feet
$K$  =  $\dfrac{dE}{dx}$  = energy loss or stopping power,
MeV per gram per $M^2$ or sq.ft. -
Mrad per minute - mA.

Energy loss K is dependent on the E energy of the electron, distance to the absorber, the density of the absorber, backscattering of electrons, and scanner window material and thickness.

It has become a standard procedure to determine K experimentally at a given voltage and under actual operating parameters.

Thus line speed can be calculated, where

$$D_r = \frac{D}{t} \tag{2}$$

Substituting into Equation (1)

$$\frac{D}{t} = K \frac{I}{X \cdot Y}$$

$$
\begin{aligned}
D &= \text{dose, Mrads} \\
t &= \text{time, minutes} \\
\text{L.S.} &= \frac{X}{t} = \text{line speed, feet per minute} \\
K &= \text{sq. ft. - Mrad per minute - mA} \\
Y &= \text{scanning width, feet}
\end{aligned}
$$

$$\text{L.S.} = K \frac{I}{D.Y.} \tag{3}$$

For a 300 kV - 200 mA system, with a 48-inch scanner, K values of 20 sq. ft. per minute-mA were measured on an Electron Processing System. The following line speeds could be achieved at the indicated dose level over four foot product widths:

| | |
|---|---|
| 2 Mrads | 500 feet per minute |
| 4 Mrads | 250 feet per minute |
| 6 Mrads | 167 feet per minute |

Radiation systems are designed in modular structure, so that additional units could be placed on a production line to double or triple line speeds with only a slight increase in line space. Thus, radiation processing speeds can be

matched to existing line speeds or to future processing
requirements.

The total radiation processing system would consist of
three items: power supply, accelerator scanner, and control
console.

The power supplies are usually oil-filled transformers
which permit line voltage to be increased and rectified to
the d.c. voltage required for the application. A high voltage cable is used to transfer the voltage from power supply
to the electron accelerator.

Picture #1 shows two 6 foot electron scanners being
powered from one power supply. These units are mounted over
a casting drum. Liquid resin was cast onto the moving drum,
irradiated and removed in sheet form from the far side. Web
handling equipment conveyed the film outside the shielded
processing area. Up to three accelerators can be powered
from one supply.

At the top of the accelerator, a cathode filament is
heated and the electrons produced are accelerated down an
evacuated tube. At the exit point, a stream of very intense
electrons emerge approximately one-half inch in diameter.
In order to utilize this source of energy, an electromagnetic
system is applied to the electrons, sweeping them back and
forth at 200 cycles per second in widths up to 6 feet.

Picture #2 shows the complete electron processing system. In this system, two accelerators are cable-connected
from one power supply. The control console is key-operated
from the outside, while within all machine parameters can be
preset prior to locking the cabinet door. The system is
self-stabilizing and automatic; thus no operator in atten-
dance is required.

In the coating field, the need has always been for
greater current requirements at only a modest increase in
the equipment price.

As the current levels are increased, from Equation (1),
dose rates are also increased. Figure 2 indicates the dose
rate effect of decreasing percent gel with increasing dose
rates. Thus, doubling the current would not result in a

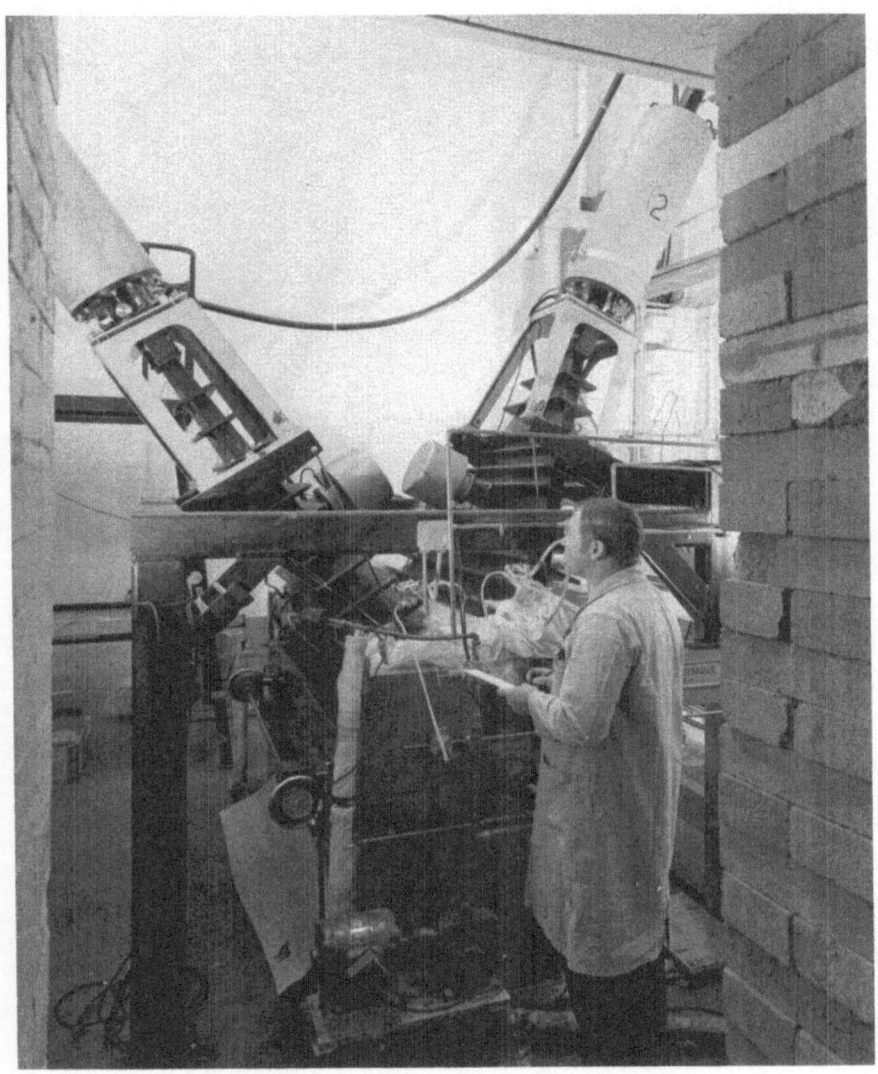

Picture No. 1

Picture No. 2

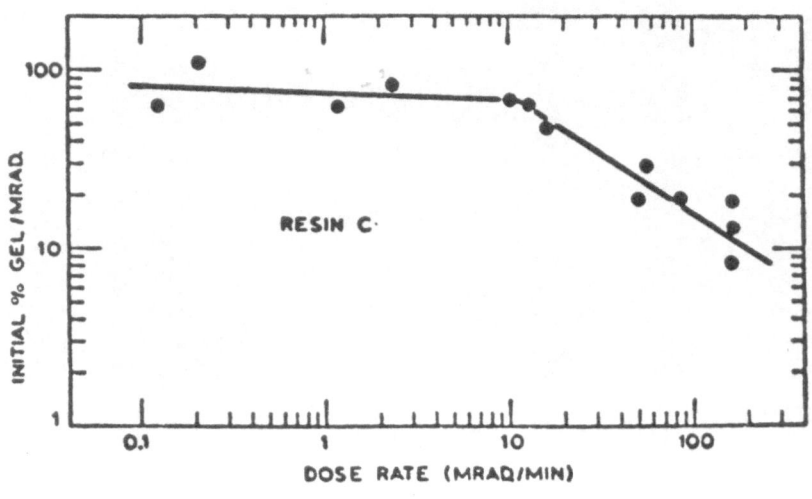

Figure 2.

two-fold increase in throughput.    In addition, increasing
the current intensity on the scanner window produces a heating problem.

In an attempt to solve both these problems, a new
scanning module was developed specifically for the coating
industry.  This unit scans in both the X and Y planes, with
the X width being 5 times larger than our standard scanner.
Thus, not only is the window heat dissipated over a larger
area, the electron area has been greatly increased; thereby
lowering the time average dose rate.

Picture #3 shows this scanner mounted beneath an electron accelerator.

Picture No. 3

Scan uniformity is very important as the product moving
through the electron cloud should receive uniform dose levels
over the entire product width.  Variation will result in
lower productivity since the minimum dose will be the determining criteria for throughput.

Figure No. 3 shows the scan "Y" profile as measured on
the area scanner under production conditions.

Figure No. 4 shows a suggested in-line installation for
the curing of coil coating.  Special mechanical interlocks
are provided in the top plate, to prevent operation of the
accelerator unless it is securely mounted into the plate
section.

During the past five years, the unit cost of processing
with radiation has been greatly reduced with the upgrading
of the equipment.

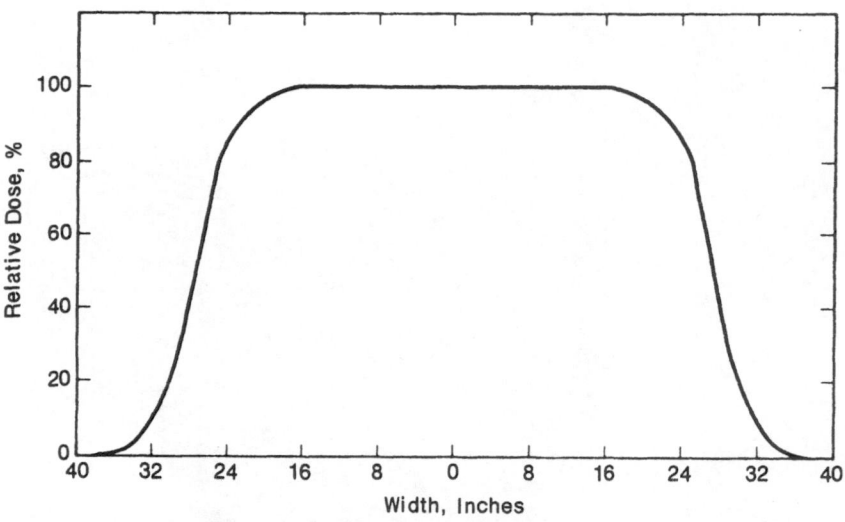

300 kV Y-Scan Profile for Irradiation of a 48-Inch Wide Web at a 6'' Air Path.

Figure 3.   Area Scanner "Y" Profile.

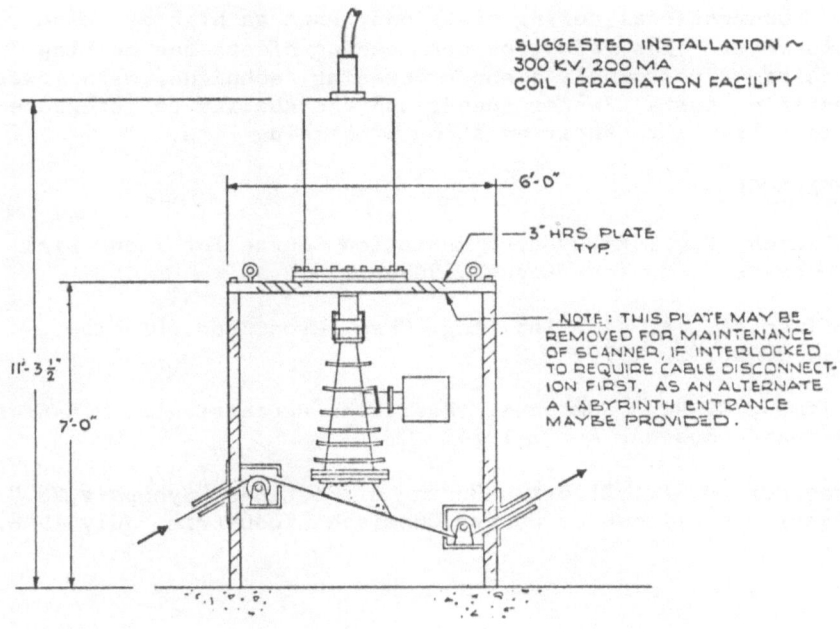

SUGGESTED INSTALLATION ~
300 KV, 200 MA
COIL IRRADIATION FACILITY

Figure 4.

The following is the estimated annual operating cost for curing, using a 300 kV - 200 mA unit with an area scanner, based on a 2-shift, 4000 hours per year operation and 4 Mrads.

|  | Cost per Hour |
|---|---|
| Amortization of Investment - 5 Year Straight Line | $10.00 |
| Utilities: Water, Power, Gas | 1.80 |
| Maintenance Service Contract by High Voltage Engineering Corporation | |
| Parts | 1.00 |
| Labor | 1.25 |
| | $14.05 |

Productivity:   250 ft./min. x 4 ft. x 60 min.
                = 60,000 sq. ft./hr.
Unit Cost:      2.34 cents per 100 sq. ft.

Conventional curing costs can range as high as twice this cost.  Thus, electron beam curing offers the coating industry a relatively clean processing technique, with lower operating costs, faster speeds and the ability to integrate this system into their existing processing line.

REFERENCES:

Rofkirch, E., IAEA, "Large Radiation Source for Industrial Processes."  Munich, August 1969.

Hoffman, A., et al., Ind. Eng. Chem. Prod. Res. Develop., Vol. 9, No. 2, 1970.

Hoffman, C. R., "Commercial Radiation Processing of Plastics Seminar", Boston, April 1970.

Trageser, D. A., Electron Curing of Coatings "Synopsis 26." Magazine published by Synres Chemische Industrie, July 1968.

# PRODUCTION SYSTEMS FOR ELECTROSTATIC SPRAY APPLICATION OF POWDER COATINGS

Hani T. Azzam, President

Interrad Corporation

1212 Post Road, Greenwich, Conn. 06878

## INTRODUCTION

Several Factors have combined to arouse interest in powder coating. Chief among these are:

1. Concern over air pollution and more stringent air pollution laws.
2. The development of new powder coatings.
3. Improved powder coating application techniques.

Of course, none of these factors would be sufficient to create the interest which we see today in powder coatings, unless powder coatings also offered performance and economic advantages as well. The superiority of powder coatings over most liquid paints is still a controversial subject, but the documentation is beginning to appear. (1, 2, 3) Also, although coating powder is more expensive than liquid paint based on straight comparison of the cost of the solids which form the finished film, the applied cost is frequently less. Powder is 100 percent solids and utilization approaches 100 percent. Losses with paint due to overspray are seldom less than 35 percent. Even liquid electrostatic paint application systems seldom provide more than 80 percent material utilization.

Powder coating also provides other economies, such as:

1.  Complete elimination of solvents or thinners, their handling and storage.
2.  Complete automation of the coating line.
3.  Elimination of paint-arrestor filters, and waterwash chemicals.
4.  Simpler clean-up and maintenance.

## ELECTROSTATIC POWDER SPRAY COATING

Of the improvement application techniques, the electrostatic spray process is by far the most significant. Although powder coatings have been on the scene for two decades, application was primarily by fluidized bed until recently. The limitations of the process restricted powder coatings to certain specific applications, primarily those involving highly corrosive service.

Only relatively thick coatings, 6-mils and up, can be applied by fluidized bed. As few coatings, either decorative or functional, are more than 3-mils thick, fluidized bed is not suitable. Other limitations include difficulty in coating large and irregularly shaped parts and parts having different cross-sections.

In contrast, electrostatic spray can apply powder coatings of 1-mil or less. The heat retention properties of the part are not a factor. The same thickness can be applied to thin as well as thick sections. Thin materials such as foils, which would be impossible to coat in a fluidized bed, can be coated electrostatically.

In the electrostatic powder spray process, powder is drawn from its container and carried to the spray gun by compressed air. Individual particles of powder are electrostatically charged as they pass through the gun. The part to be coated is grounded and thus at a lower potential than the charged particles, so an electrostatic field is generated between the tip of the gun and the workpiece. Particles projected from the gun are attracted

to the surface of the part and adhere to it until they are fused
to the surface and heat cured in  the bake oven into a
homogeneous coating.

## ELECTROSTATIC SPRAY EQUIPMENT

The essential components of an electrostatic powder spray
unit are:

1.  A powder container.
2.  A system for conveying metered amounts of powder from
    the container to the gun.
3.  An electrical system for developing the high voltage charge
    which will be imparted to the powder particles.
4.  A powder spray gun.

The powder container is simply a hopper equipped with some
type of mechanism for keeping the powder free flowing and
preventing bridging. The simplest mechanism for this is a motor
driven agitator. Powder is withdrawn from the bottom of the
container hopper,  usually by a venturi in the compressed air
line with  air pressure controlling the rate of powder feed.

The high voltage generator increases line voltage to the high
voltage required to charge the powder. Most employ a cascade
generator developing up to 70 KV and more. The cascade
generator produces high frequency pulses of direct current to
establish a strong electrostatic field at the tip of the gun and
between the gun and the workpiece. Although the voltage
developed is high, current is measured in fractions of a
milliamp.

Gun barrels are usually of an insulating plastic, frequently
nylon for maximum abrasion resistance. In many guns the powder
articles are charged only as they leave the gun and enter the
electrostatic field between the tip of the gun and the workpiece.
In some guns the particles are also charged within the gun,
increasing charging efficiency.

Fig. 1 - 5 gun, 2 color unit

Guns are equipped with some sort of adjustable deflector at the nozzle to permit the spray pattern to be varied. Most deflectors are stationary and are adjusted by varying distance between the nozzle and the deflector, or by changing the deflector. Some deflectors rotate during operation, but this is not essential and adds another moving part to the system.

The design of the basic equipment varies. In some the hopper and high voltage generator are an integral unit with all controls located conveniently together. In others the hopper and high voltage unit are separate, each with its own controls.

In some multi-gun systems all guns receive powder from a common source having but one voltage output, and are fed at identical rates from a single powder container. In others, each gun has its own power source with its own electrical controls for voltage and compressed air controls for powder feed rate, even though all guns are fed from a common hopper. (Fig. 1) The latter system afford maximum flexibility as the voltage and powder feed rate of each gun can be controlled individually for optimum deposition efficiency. Also, failure of one gun does not knock out the entire system.

The high voltage required depends on the powder being sprayed and generally ranges from about 35 to 75 KV. For example, thermosetting polyester powders insure a better coating when charged at 40 KV or less. Epoxy powders perform well in the range from 40 to 70 KV. Vinyl powder generally require a higher charge, on the order of 60 to 75 KV. These are the voltages required at the tip of the gun where charging occurs.

## AUTOMATIC SYSTEMS

One of the major advantages of electrostatic powder spray coating is the ability to completely automate the system. (Fig. 2) Such systems are highly efficient. They can operate virtually unattended with very little maintenance.

Fig. 2 – Automatic Powder Coating System

Figure 3 is a schematic of an actual epoxy powder coating system for applying two colors. The use of duplicate spray booths and recovery systems for each color is one of the ways to provide for color change.

## PRETREATMENT

As in a liquid paint line, the extent of surface pretreatment for a powder coating line depends upon the type of powder used and the nature of the application. The one essential requirement is that the surface be clean, free of dirt, oil grease, oxide coatings and other contaiminants. This is usually accomplished by vapor degreasing or an alkali wash. Phosphatizing improves adhesion of powder coatings. For maximum protection against extremely aggressive environments, grit blasting of the surface will further improve the bond between a powder coating and the surface. Most thermoplastic powders require a liquid primer applied over the pretreated surface.

## COATING SECTION

Powder coating booths are almost completely enclosed. Openings are provided only for guns, for entry and exit of the parts and for powder recovery. Maximum enclosure improves powder recovery efficiency and system cleanliness.

Booth configuration is primarily a function of conveyor speed, number and location of guns, rate of powder application and size and shape of the parts to be coated. Conveyor speed and powder output govern coating thickness.

Usually it is better to use several guns at moderate output rather than to force high output from a few guns. Lower outputs promote higher deposition efficiency and reduce overspray. In addition, high outputs may tend to cause the guns to clog and require more frequent cleaning. Effective powder delivery is limited by the ability to charge all of the particles passing through the gun. Uncharged particles will not be attracted to

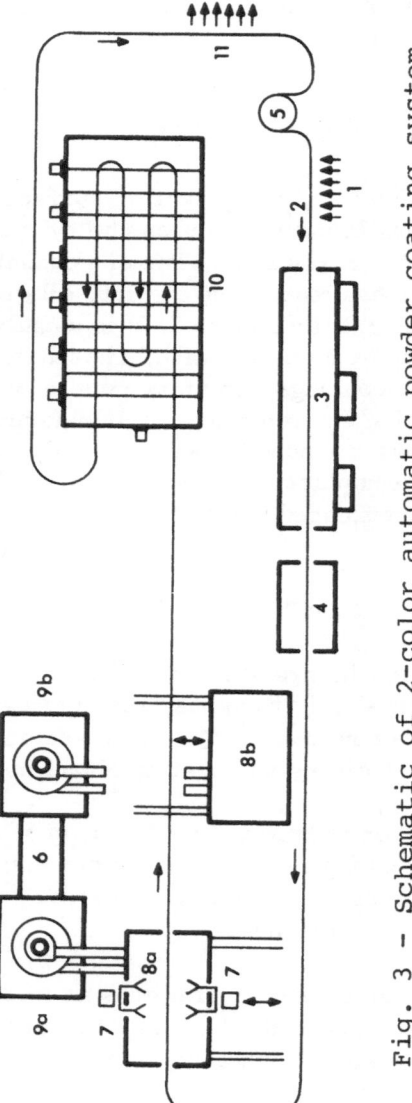

Fig. 3 – Schematic of 2-color automatic powder coating system

the workpiece. Individual control of voltage, air pressure and powder feed rate to each gun provides maximum flexibility in designing a system for optimum results and economical powder consumption.

Color change can be accomplished in several ways. As in Figure 3, each color can have its own spray booth and recovery system. These booths are on rails and can be moved quickly into, or out of, position. Another method is to design the booth and recovery system to be easily cleaned when changing colors. A third technique is to collect the overspray of all colors in one recovery system. The resulting powder mixture is then used for functional, rather than decorative, coatings.

## SPRAY BOOTH DESIGN

The general configuration of the spray booth is determined by the size of the part, the speed of the conveyor and the number of guns and their arrangement. Guns may be mounted on one side, or on both sides.

To minimize escape of powder from the booth and to provide maximum control of air flow into the booth, openings for parts to be coated should be as small as possible. Typical openings are 6 in. larger than the part on all sides. If different size parts are to be coated in one booth, openings should be adjustable.

Booth sidewalls are usually vertical. (Fig. 4) The floor of the booth should slope toward the collection ducts to minimize the tendency of powder to cling to the surface. All interior surfaces should be smooth and free of projections or ridges on which powder might collect.

## BOOTH AIR FLOW

The most critical aspect of booth design is to provide sufficient air flow. One of the most common mistakes in early powder

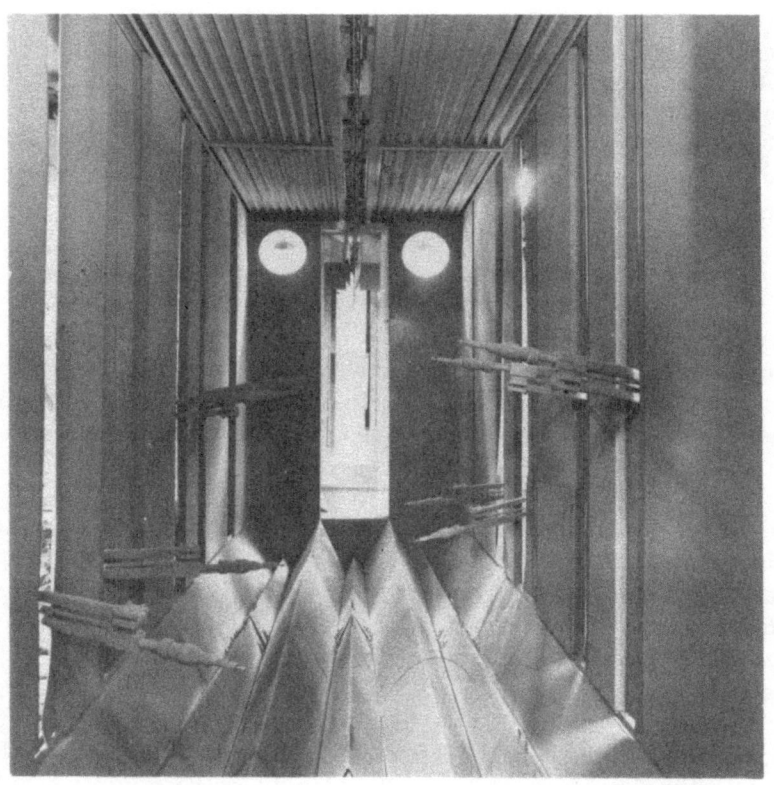

Fig 4 - Interior of Automatic Powder Coating
System

coating systems was failure to provide sufficient air flow at proper velocities.

There are three primary requirements for booth and recovery system airflow:

1. Powder concentration should never exceed a safe level. At the present time in the United States this level is 10 oz per 1,000 cu ft air.
2. Intake air velocity at booth openings should be at least 80 fpm to keep powder in the booth. But, it should not exceed 100 fpm at booth entrance or exit so as not to blow powder from the parts.
3. Within the booth, vertical air velocity should not exceed 75 fpm. Higher velocities may disrupt the powder pattern and reduce deposition efficiency.

The amount of air required to maintain a safe powder-air concentration depends on the number of guns and their maximum output. For instance, if 5 guns are required and each gun has a maximum output of 1 lb per minute, the maximum powder output of the system is 5 lb or 80 oz per minute. Therefore 8,000 cfm of air must be exhausted from the booth to maintain a safe concentration.

Plant air is drawn into the booth through the various booth openings, entrance and exit, ceiling slot and gun openings. The volume of air which can enter the booth through these openings is limited by the requirement that air velocity at booth openings be more than 80 but less than 125 cfm.

If necessary, additional air can be introduced through a plenum chamber in the top of the booth, with baffles controlling air velocity from the plenum. (Fig. 4) The best source of this air is the recovery system exhaust. However, regulations in some localities may prohibit its use. Also, reuse of recovery system exhaust is impractical if the same booth and recovery system are to be used for several colors, as it makes the booth difficult to clean. In such cases the required additional air is drawn from the plant.

The third air flow requirement insures that the downward flow of air through the booth does not interfere with spray patterns or blow powder from the parts.

## RECOVERY SYSTEM

Efficient recovery of oversprayed powder is the key to economical powder coating. Basically recovery systems consist of one or more cyclones or bag filters, or combination of both.

Bag filters are the most efficient in terms of powder recovery. They can collect nearly 100 percent of overspray. However, bag filters capable of handling the large air volumes involved must be very large. Also, bag filters are relatively difficult to clean and a separate filter must be used for each color.

Cyclones are more practical for multi-color operations. (Fig. 5) High efficiency cyclones are capable of recovering from 95 to 98 percent of oversprayed powder depending upon the particle size. They are relatively easy to clean, and the same unit can be used for more than one color.

Cyclones are frequently followed by bag filters to remove fines. This is essential if the exhaust air is being returned to the plant rather than vented to the atmosphere, and in some cases an absolute filter may be required to remove any sub-micron particles which may pass through a bag filter.

Powder collected in both bag filters and cyclones can be returned to the powder supply hopper automatically. The return system should include a screen to reject oversize particles, agglomerates or any foreign matter which may enter the system.

Recovery system exhaust air which is not fed back to the booth can be vented inside the plant, if local regulations permit, or outside to the atmosphere. Obviously it is desirable to exhaust as little as possible outside to minimize heat loss. This practice is economical even if it calls for an extra filtering step.

Fig. 5 - Powder recovery cyclones
and bag filter

## CURING

Powder coating lines and ovens require no flash-off space. The coated part proceeds immediately from the spray booth to the bake oven. Oven temperatures and curing times depend on the part and the powder. Temperature of the surface of the part should be kept below the decomposition temperature of the powder to avoid the production of flammable fumes.

## POWDER REQUIREMENTS

Indications are that, as a finishing process, electrostatic powder spray coating will soon far outstrip fluidized bed powder coating and will even surpass electropainting within a few years. The trend in the United States is not only toward electrostatic powder spray, but toward automatic application, as well. This means not only automatic spraying of the parts being coated, but automatic recovery and return of oversprayed powder. Powder manufacturers should realize that this imposes much more severe conditions than manual operation in the laboratory.

Spherical powder particles accept and retain an electrostatic charge better than other configurations. For electrostatic spray the optimum particle size range appears to be between 40 and 60 microns. Larger and smaller particles do not adhere as well to the surface of the part being coated, and, as a result after several hours of operation in a recirculating system, recovered overspray tends to contain too high a percentage of fines and oversize particles. The result is a change in the quality of the surface finish of the coating for no apparent reason. Improper particle size distribution is a major cause of orange peel.

Production of powders containing only 40 to 60 micron particles however desirable, is impractical at this time. The usual solution is to add virgin powder along with the recovered overspray. Equipment is available to continuously replenish the recirculated overspray with new powder at a predetermined rate.

Another problem is absorption of moisture. If the powder absorbs moisture too readily, it will tend to agglomerate and cake. The difficulty this can create in a recirculating system should be obvious. The trend in the United States appears to be toward melt mixed powders rather than dry blended products. especially for use in automatic systems.

## WHAT'S NEEDED

The most urgent needs of the powder coating industry today is the development of standards -- safety standards and quality standards.

Powder coating lines are less hazardous than solvent paint lines. The energy levels required to ignite a dust-air mixture are much greater than those which will ignite normal paint solvents. Nevertheless, a hazard does exist and must be taken into account in the design of a powder coating system. However, because there are no recognized standards in this country, equipment must be designed for maximum safety. As a result, systems are more costly to install and to operate than they need be.

What is needed is a testing and certification program to determine the realistic safety requirements for various types of powders. Such standards will result in more economical operation.

Finally, speaking as an equipment manufacturer, there is an urgent need for the development of quality control procedures to assure the equivalent quality of powders from different manufacturers and between lots from the same manufacturer. Hopefully such procedures will be simple so that they can be used for incoming inspection to give assurance of particle size distribution, color match and gloss consistency. At the present time, the first time a user discovers that he has received a substandard powder is when he observes a poor coating on the finished product.

Powder coating is here to stay, and most coating powders will be applied by automatic electrostatic spray for the foreseeable future. However, like all new processes, powder coating will have its growing pains. These can be minimized if the lines of communication between the end user, the powder supplier and the equipment manufacturer remain open and all three groups work closely together.

## REFERENCES

1.  "First High Volume Epoxy Powder Coating Line in Auto Industry," INDUSTRIAL FINISHING, August 1971, pp. 20-23.

2.  Breton, "Experience with Powder Coating," PROCEEDINGS FIRST NORTH AMERICAN CONFERENCE ON POWDER COATING, February 1971, McLean Hunter Ltd, pp. 85-89.

3.  Denice, N.J., "Electrostatic Powder Line Produces Better Coatings at Lower Cost," INDUSTRIAL FINISHING, January 1972.

4.  Kunz, W., "Considerations in the Design of Automatic Electrostatic Spray Coating Installations," PROCEEDINGS FIRST NORTH AMERICAN CONFERENCE ON POWDER COATING, February 1971, McLean Hunter Ltd. pp. 59-63.

5.  Vork, W.D., "Electrostatic Powder Deposition and Recovery Systems," PROCEEDINGS FIRST NORTH AMERICAN CONFERENCE ON POWDER COATING, McLean Hunter Ltd., pp. 55-58.

6.  Strobel, R.F., "Safety Considerations with Powder Coating," PROCEEDINGS FIRST NORTH AMERICAN CONFERENCE ON POWDER COATING, February 1971, McLean Hunter Ltd. pp. 105-109.

7.  Kut, S., "Safety Considerations in the Use of Powder Coatings," POWDER COATINGS, 1970.

8.  Azzam, H.T., "Coatings without Solvents," MACHINE DESIGN, March 18, 1971, pp. 91-95.

9.  Smarsh, J., "Powder Coating - How, Why, When," 49th Annual Meeting of Federation of Societies for Paint Technology, October 1971.

10. Hoefler, H. and Peter, E.,"Designing Automatic Electrostatic Powder Spray Systems," PRODUCTS FINISHING March 1972

11. Azzam, H.T., "Application Techniques for Powder Coating,"
     PAINT & VARNISH  PRODUCTION, April 1972.

# ELECTROSTATIC POWDER COATING - POTENTIALLY A POLLUTION FREE FINISHING METHOD

Emery P. Miller

Ransburg Corporation

Indianapolis, Indiana   46208

A certain amount of waste material has always been associated with any finishing operation. Just exactly how great this effluent is, the exact nature of its composition and one method of avoiding it is of interest to us in this discussion.

To apply a finish to an article, it is usual to dissolve a resin film-former in a suitable organic solvent so that it can be spread upon the surface to form the desired solid coating. In the spreading process, regardless of its nature, there is always some material which escapes deposition.

In the standard air spray application process, it is not unusual for this loss to be as high as 55 to 60 percent of the material sprayed at the object. This lost material must somehow be captured and retained if it is not to be distributed about the area to accumulate on parked cars and other adjacent structures. In addition, because the applied liquid coating is only half film-former and the other half is a solvent which must evaporate, there is a considerable built-in loss of this volatile component. This volatile component leaves the material from the time it is first atomized until it is finally cured at the end of whatever oven facility may be used.

If we examine this situation as it refers to a specific operation, we can readily grasp the potential magnitude of such operations as pollution sources. Consider a manufacturing operation in which 2400 automatic washing machine cabinets are being air sprayed in every 8-hour period. In such an operation, approximately 300 gallons of material -- 8 cabinets per gallon -- will be used. The deposition efficiency will be about 45 percent. This means that 135 gallons of the material sprayed will be collected on the parts while the other 165 gallons must be considered overspray waste to be collected. Assuming the material as sprayed is 40 percent solids by volume, we need to capture 66 gallons of solids and 99 gallons of solvent which has evaporated to a vapor.

Further the material deposited on the work was likewise 60 percent solvent, so we need to capture another 81 gallons of solvent in vapor form from this source.

These materials originate in different areas of the operation. The 66 gallons of oversprayed solids and 99 gallons of oversprayed solvents originate in the spray booth at the point of application. The other 81 gallons of solvent originate at all points between the spray application position and the end of the curing oven.

The finishing man for years has attempted to cope with these effluent materials because they have been troublesome and represent dollars lost to the operation. His efforts have been made mainly on the basis that the loss was an inherent part of the process, and he assumed that the best that could be done was to capture the waste materials. On this basis, elaborate water wash spray booths have been installed to capture the oversprayed solids and wash the solvents generated in the spray zone from the exhaust air. Sludge was dutifully collected from the water sump of the booth and disposed of in adjacent land fill, burned where possible, or in some cases, recycled as second grade coating.

Little or no attempt was made to control the solvent effluent from ovens. These generally were pumped into the atmosphere with the hope that ultimate dilution would be sufficiently great that they would pass neighborhood visual and nasal inspection.

One effort was made to improve the overall situation. Urged by the basic economic reasoning of curtailing the loss of the rather expensive coating material, electrostatic liquid spray methods were introduced where applicable, to replace the rather inefficient air spray gun. With these methods, deposition efficiency could be as high as 90 to 95 percent on items such as the jackets considered above. This obviously reduced the overall spray losses originating at the application position, improved the economy of the operation and, in general, was a marked advance in improving the overall situation. In the above jacket operation, the same coating could be applied with electrostatic means by spraying only 145 gallons of material -- a reduction in the daily use of material of 155 gallons. Of the 145 gallons sprayed, 135 gallons would still go to form the coating on the jackets but the collection booth now only needs to handle the 10 gallons of oversprayed material -- 4 gallons solids and 6 gallons of solvents. This is a much easier task.

However, electrostatics cannot improve the situation related to the solvent vapors originating from the curing film -- the 81 gallons of solvent in liquid form still is emitted from the oven as vapor.

Afterburners were constructed in the oven stacks to reduce these vapors by combustion to acceptable non-polluting materials.

It was also conceived that the system could be lived with if the normally used solvents were to be replaced with solvents which, of themselves, caused no difficulty when exhausted into the air. This, so called, exempt solvents approach presents an

interim solution but leaves something to be
desired.

Still another approach was to attempt to formulate suitable resin coating systems which were
water based and used at best only a very small
portion of organic solvents. This would result in
only water vapor being emitted into the air and
certainly should be acceptable. This approach, at
present in the development stage, has shown some
promise as of this date.

Another general approach is to visualize that,
somehow, all solvents could be removed from a
liquid coating system. A possible approach to this
would be through the use of composite liquid systems
which would polymerize after application to a solid
film without residue. Such a system is a possibility. It still needs to be worked out.

The dry powder application method of coating a
part presents still another possibility of utilizing
a solvent free coating system. In this process, the
particular resin material to be applied is prepared
in the form of a dry dust-like powder. It is deposited on the surface to be coated and is fused there
to form the desired coating. In one powder coating
method, the fluidized bed method, the powder is
placed in a container having a porous bottom. Air
is pumped upward through this bottom and the powder
to bulk or fluidize it.

The part to be coated is preheated to a temperature above the fusion temperature of the powder,
and then it is immersed into the fluidized bed. The
powder, upon contacting the hot surface, melts and
fuses. After an appropriate time, during which the
desired layer forms by fusion on the part, the part
is withdrawn. Subsequent flow of the resin takes
place either due to retained heat, or the part can
be post heated. This method has received rather
wide acceptance. With it, quite thick films of
coatings can be formed on many types of parts, but
the variation in film thickness produced and the
careful control required to insure consistent
results have limited its application. It does,

however, permit the formation of coatings without
producing polluting residues after the manner of
liquid coating operations.

In another powder method, the powdered material
is actually sprayed at, or distributed over, the
surface to be coated using a suitable spray gun.
The part is again heated so the contacting powder
will melt and adhere to the surface. With this
method, which is very similar to the dry porcelain
enameling method, the resultant film can be very
irregular in thickness and dependent upon the heat
retained by various portions of the part. Casting
portions will retain heat longer than adjacent sheet
metal portions and, so, will receive a different
coating. The material sprayed past the object must,
indeed, be recovered to prevent pollution and to
make the process economical. No solvents, however,
are available to be released.

The most recently introduced of the powder
methods, the electrostatic powder method, is proving
to be the most generally useful. In this method,
the distributing device or spray gun is associated
with a voltage supply so that the powder leaving the
gun is electrostatically charged much as in the case
of the electrostatic liquid system. The article
on the conveyor is electrically grounded. An
attraction exists between the charged powder particles of one electrical sign and the article at the
other, so the particles move toward and accumulate
on the article surface. The article can be at room
temperature, or it can be heated. When heated, the
particles will be sintered upon contacting the hot
surface and will adhere. Material will continue to
accumulate from the spray so long as heat is available to produce sintering. The part can be post
heated to cause the powder to further fuse and flow.
Thick films can be produced when parts are
preheated.

In another electrostatic powder spraying modification, the parts are not preheated but are coated
while at room temperature. In this case, the
charged powder particles attracted to the parts
surface will collect there being held in place by

their electrical charge. Because these particles
are of insulating material, only part of their
charge flows off to the grounded part. As the powder layer accumulates, the electrical charge on the
surface accumulates. Soon the residual surface
charge will preclude any additional powder attraction. At that point, the maximum powder film has
been collected. This usually occurs with most normal powders at a thickness which will fuse and flow
to about 1 to 1.5 mils. It is interesting to note
that, although the charge accumulation limits the
film thickness that can be applied, it also promotes
film uniformity over the part surface.

In the electrostatic powder method, not all
powder sprayed at the object is accumulated on the
part. There is some over-sprayed powder which must
be trapped. By way of example, we might consider
the consequences of doing, with powder, the same
finishing operation that was considered above with
liquids. With a reasonable good powder, a finish
film 1.0 mil thick can be obtained when 150 square
feet of surface is uniformly covered by one pound
of powder. To obtain the required 1.5 mil film,
a coverage of 100 square feet per pound could be
used. To produce the finish desired on the washer
jackets described above -- 39,000 square feet in 8
hours at 1.5 mils -- would require that 390 pounds
of powder be applied per shift.

Powder can be electrostatically sprayed at a
transfer efficiency of about 80 percent. It would,
therefore, be necessary to actually spray 490 pounds
of charged powder at these parts to perform the
needed coating operation. We would, thus, have
approximately 100 pounds of powder oversprayed in
8 hours. This would have to be trapped for re-use
because it would not be permissible to allow this
amount of powder to escape into the atmosphere.
Recovery is also desirable to increase the efficiency of the process.

The operation must, therefore, be carried out
in a ventilated enclosure so as to confine the oversprayed material. Such powders are potentially
hazardous from an explosive viewpoint. The atmos-

phere in the enclosure and recovery system must be
held below the lower explosive limit of the particular powder material.  Tests have shown that the
lower explosive limit of most plastic powders is
near .02 ounces per cubic foot of air or 20 grams
per cubic meter.  This corresponds to mixing 800
cubic feet of air with each pound of powder.  If we
apply this condition to the above operation, we can
obtain fire safety by introducing into the spraying
enclosure with the 490 pounds of powder 392,000
cubic feet or 11,000 cubic meters of air during the
8-hour period.  This corresponds to about 1,000
cubic feet per minute introduced into the enclosure
-- a perfectly reasonable value.

The 100 pounds of oversprayed material must be
handled in some manner.  In electrostatic spraying
operations, this material is usually carried from
the spray enclosure by the ventilation air.  It is
carried to an attached combination cyclone-bag type
air cleaner through which the air is drawn.  Such a
cyclone cleaner will remove an estimated 80 percent
of material of the size of these powders.  This
means that of the 100 pounds of material oversprayed
in the above process, 80 pounds will be recovered in
the 8-hour period.  The follow-up bag filter is
estimated to be 99 percent efficient so it will
capture roughly, 19.8 pounds of the remaining 20
pounds of material.  This leaves 0.2 pounds or 90.8
grams to be distributed in the exhausted air volume
of 11,000 cubic meters.  On the average, therefore,
the exhausted air will contain about 9 milligrams of
powder per cubic meter.

The upper toleration limit for random dusts is
stated to be 15 milligrams per cubic meter, so the
exhaust air from the operation after passing the
cleaners can safely be reintroduced into the spray
area or exhausted to the out-of-doors as desired.

From this calculation of a typical coating
situation, it is evident that electrostatic powder
coating methods can be safely used to meet the
pollution standards for operations of this type
when designed to incorporate correct spraying
enclosures, powder recovery means and efficient

application devices.  The reclaim of powder, if
collected from a single color operation, can be
reconditioned without difficulty and recycled at
once to the powder supply.  By such re-use techniques, the overall efficiency of application can
be as high as 99 percent.

Since there are no solvents available in the
system, the possibility of air contamination from
this source is obviously zero.  Adequate films for
most purposes can be built up withoutdifficulty.
The process meets all the requirements of a pollution-free system.

In view of the mentioned advantageous features
which the electrostatic powder system is said to
possess, it would appear that it should readily be
adopted by all who wish to solve their emission
problems.  However, the method, which appears at
first blush to be very simple, is not without
present limitations.

Not all types of resin materials are readily
available in powder form to be applied by this
technique.  Rather severe limitations, such as
correct electrical characteristics, best particle
size distribution, best particle shape, narrow melt
range, and low melt temperature are imposed on the
ultimate material if it is to work most effectively.
In addition, these characteristics should be reproducible from batch to batch.  These requirements
have caused the powder manufacturer some production difficulties when added on top of the general
physical and chemical requirements placed on the
material by its end use.

Epoxy powders are most widely used at present
because they seem to possess, or can have built
into them most easily, these needed characteristics.
Nylon powders have, also, had reasonable usage as
has cellulose acetate butyrate.  These latter
materials require that a primer be applied to the
surface before full adhesion can be developed.
Other resins, such as acrylics and polyesters, are
being produced and are available in limited
quantities.  These, undoubtedly, will be improved

and more widely applied as experience with their
use is broadened. Polyethylene and polypropylene
powders are available, but specific uses for these
thermo plastic materials are still being developed.
As powders are further improved and new modifica-
tions introduced, the range of applications of the
method will broaden, but at the present, powder
development and application move forward hand-in-
hand at a relatively slow pace dictated by other
features of the method. The use of a particular
powder can only be proven when such a powder is
available while paradoxically the powder manufacturer is rightly hesitant to spend development
money on a powder until its use in rather large
volume is assured.

To adapt the electrostatic powder spray system
to an existing liquid line requires a major modification in much of the present facilities. Users
are holding fast to what they have with the hope
that some more readily adaptable solution to their
pollution ills will present itself. For example,
the broad use of the powder coating method is
being held up because of the developments in the
"water based" materials program.

Users who finish products in more than one col-
or hesitate at adoption because no ready method of
reclaim and re-use has been developed. To economically apply powders, it becomes imperative to
recycle that powder which escapes deposition at
first application. This means that, if blue powder
is being applied, blue powder must be reclaimed. A
shift to applying red powder and the corresponding
recovery of red powder only can be accomplished if
all equipment is cleaned at the time of change. No
simpler method has been developed short of duplica-
tion of equipment for various colors. This has been
a major limitation to the use of the system on such
applications.

Although it is not always easy to obtain the
quality of finish required for a particular application, the appearance required on many products can
be readily obtained. These situations where
satisfactory finish is obtainable should be ready

applications for the method.  Other techniques such
as thermal reflow after wet sanding are being considered and, if successful, will, undoubtedly,
broaden the application field even further.

When these apparent limitations are balanced
against the good features of the system, it would
appear that there is a plus in favor of the method.
Development effort, both in materials and process
equipment, is bound to produce results which will
add to its advantages and ease of adoption.  Each
step forward makes the method more appealing and
economical compared to alternatives.  However, as
the alternatives are improved, the position of
powder coating is further tested.  Although it is
not anticipated that powder coating will rapidly
replace all present liquid systems, it is a very
effective way of ridding a finishing operation of
its inherent pollution difficulties.  On this
basis alone, it is anticipated that powder methods
will win their fair share of the applications.

# MEASUREMENT OF OPACITY AND COVERAGE IN THIN-FILM POWDER COATINGS

Douglas S. Richart
Kenneth W. Gray
The Polymer Corporation
Reading, Pennsylvania   19603

## INTRODUCTION

Until recently, the field of powder coating has occupied a special area in the general field of plastics technology, somewhat apart and removed from conventional paint technology.  This is mainly because coatings applied from powder are normally 10 mils thick or greater and used in applications where their outstanding corrosion protection, insulating characteristics or other functional properties are important.  However, in the last several years, with the improvements made in both powder coating formulations and application equipment, coatings can readily be applied in the range of 1-3 mils.  This brings the powder coating field directly in the realm of paint technology.  While many of the tests used for conventional paints can also be used to measure films applied from powder, there are a number of other characteristics of a powder applied system which cannot be evaluated using conventional techniques.

While the measurement of opacity and coverage is an accepted procedure in the paint industry, it has not yet gained widespread use in the powder coating industry.  This is due, in part, to the relative newness of powder technology and also to the lack of reproducible procedures to quantitatively define the parameters involved.  With minor modifi-

cations, the same procedures developed for conventional
coatings can also be used with powder coatings.(1)


## THEORY

Concepts relating to hiding power and opacity may be
well-known to those in the paint industry.  However, for
those in the powder coating industry who may not be too
familiar with all the terms the following definitions are in
order:

Coating System - the combination of a coating
    material applied to a substrate.

Opacity - the ability of a coating system to
    hide the substrate.

Reflectance - the ratio of light diffused by
    a coating system to the light diffused by a
    near-perfect diffuse reflector such as
    magnesium oxide.

Scattering - the cumulative effect of reflection,
    diffraction and refraction from a coating
    system on the incident light.  The value is
    primarily dependent on the type and concentration
    of pigment.

Absorbance - that part of incident light not
    scattered by a coating system.

Coverage - the area which can be hidden by a
    coating system at a specified contrast ratio
    per unit weight or volume of coating material.

The ratio of light absorbed to light scattered, K/S, is
dependent upon the specific pigments and their concentration
in the medium.(2)  Scattering efficiency of a pigment can be
measured according to the well-known Kubelka-Monk equations:
(3)

$$\frac{K}{S} = \frac{(1 - R_\infty)^2}{2 R_\infty} \qquad\qquad \text{Equation 1}$$

$$S = \frac{2.303}{d} \cdot \frac{R_\infty}{1-R_\infty^2} \cdot \log \frac{R_\infty(1-R_0 R_\infty)}{R_\infty - R_0} \qquad \text{Equation 2}$$

where:

$R_\infty$ =diffuse reflectance for the coating system at infinite thickness such that a further increase in thickness does not appreciably affect the K/S ratio.

$R_0$ =diffuse reflectance for the coating system over a black substrate of which Reflectance $\rightarrow 0$.

$d$ =film thickness

Consequently, a method is needed to measure diffuse Reflectance, R, and the scattering coefficient, S. Diffuse reflectance can be easily measured with any number of color-imeters or spectrophotometers. To determine the scattering coefficient, S, the coating is applied over a white substrate (Reflectance about 80%) at sufficient thickness to hide the substrate and over a black substrate (%R$\rightarrow$0) at the normal thickness range. The Reflectance is measured for each versus a near-white diffuse reflector. The ratio of Reflectance over black ($R_0$) to reflectance over white ($R_\infty$) is known as the Contrast Ratio. It should be noted that $R_0$ is approx-imately equal to $R_B$ and $R_\infty$ is approximately equal to $R_w$. Using graphs derived from Equation 2, and the measured contrast ratio, the scattering coefficient, S, can be determined.

### PROCEDURE

In order to determine the contrast ratio and subsequently the scattering coefficient, two types of substrate are required:

White - flat white glass plate which is opaque and reflects 80% of a light striking the surface can be used.

Black - black glass or conventional paint test panels coated with a flat black lacquer (R$\rightarrow$0) can be used. Reflectance level on a conventional colorimeter (Y tristimulus value) should not exceed 3% versus barium sulfate.

For each coating system to be analyzed, one white glass panel and three flat black panels are required. After the panels have been prepared they are weighed to the nearest milligram and the weights recorded. The powder coating is then applied keeping the back of the panel covered to allow powder deposition only on the front of the panel.

## WHITE GLASS PANEL

Spray one panel to a depth of 4 mils or greater keeping the coating as uniform as possible. Fuse the coating long enough to achieve full fusion and flow of the film. If care is taken, this point can be determined before the film is fully cured and the coating can be removed from the substrate thus allowing multiple use of the plate glass panel. After cooling to room temperature the panel is reweighed and the coating weight determined.

## BLACK LACQUERED PANEL

The three panels are carefully sprayed so that the thickness of the electrostatically applied coating when fused is in the normal range (1.5–3 mils). Powder deposition on the back side of the panels can be avoided by placing the panel against a flat, conductive, grounded surface.

The coated panel is placed in an oven and the powder is fused to the finished film. In this case, the normal coating cycle can be used since the metal panels cannot be reused.

The panels are carefully reweighed and the coating weight determined.

The coating thickness can then be determined based on the weight and specific gravity of the film:

$$X(mils) = \frac{(1000)(film\ weight\ in\ grams)}{(surface\ area,\ in^2)(2.54)^3(sp.gr.)} \qquad \text{Equation 3}$$

The Reflectance is measured and recorded for each of four specimens versus barium sulfate or magnesium oxide using the Y tristimulus filter of a colorimeter or by the weighted ordinate method in conjunction with a spectrophotometric curve.(4) The contrast ratio (CR) is then calculated for each black panel using the following equation:

$$\text{CONTRAST RATIO(CR)} = \frac{R_o}{R_\infty} \approx \frac{R_B}{R_W} \qquad \text{Equation 4}$$

After the Contrast Ratios for each black panel have been calculated, their respective Scattering Coefficients can be determined. These may be calculated using Equation 2 or, more easily, determined through the use of the same equations in the form of a graph.(5)  (See Figure 1.)  Scales provided on Figure 1 include:

1. Reflectance over black – $R_o$ or $R_B$
2. Contrast Ratio – CR
3. Reflectance over white – $R_\infty$ or $R_W$
4. Scattering Power – SX

Any coordinates can be used for determining the scattering power, SX, but $R_o$ and CR will be used in this example. Whatever values are used to determine SX, whether $R_o$ and CR, $R_o$ and $R_\infty$, or CR and $R_\infty$, the same should be used for all determinations. Small differences arise due to nonideal coating surface characteristics which can introduce an error if the same procedure is not followed in all cases. (6)

Determination of the SX value is made by locating the horizontal line representing $R_o$ on Figure 1 and the intersection of that line with the calculated contrast ratio. That point is interpolated using the near-vertical SX lines and recorded for each black panel.

After the Scattering Power, SX, has been determined, the scattering Coefficient is easily determined as follows:

$$S = \frac{SX}{X} \qquad \text{Equation 5}$$

(Units are $mil^{-1}$)

From the values of $R_\infty$ or $R_W$ and S, the expected coverage can be calculated for specified Contrast Ratios. Complete hiding of the substrate, for all practical purposes, is at CR=0.98, but often a lower degree of hiding is satisfactory. (i.e., CR=0.95 or CR=0.93. Previous publications (1,7) provide equations and tables relating to hiding ability as a function of $R_\infty$ or $R_W$. Factor "A", defined in these publications, when multiplied by the scattering coefficient, S, show the number of square feet coverable at a specified

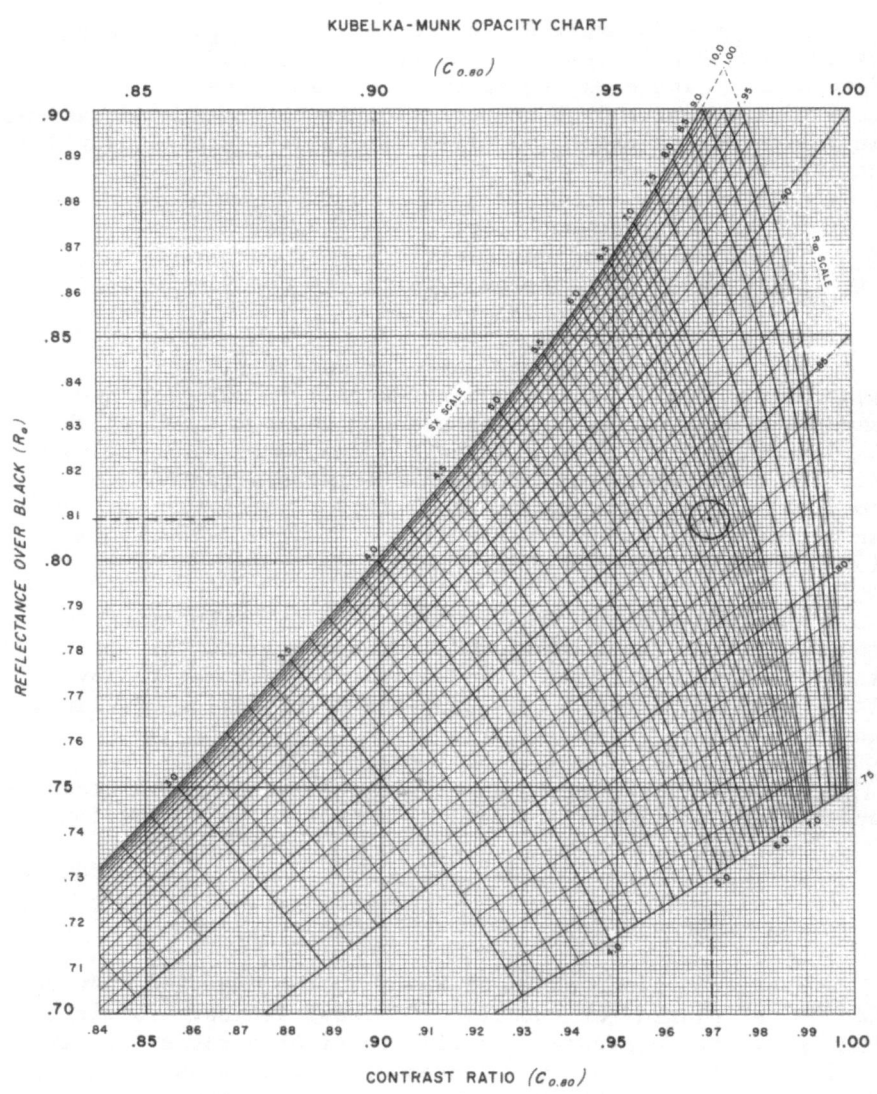

Figure 1

contrast ratio.  Factor "A" is defined as:

$$\text{Factor "A"} = \frac{(1604.17)(b)}{\coth^{-1}\left\{ \dfrac{[(a+d)^2 - 1.0/CR]^{\frac{1}{2}} + d}{b} \right\}}$$    Equation 6

where:  $a = \frac{1}{2} \left( \frac{1}{R_\infty} + R_o \right)$    Equation 7

$b = a - R_\infty$    Equation 8

$d = (1.0 - CR) / (2)(R_s)(CR)$    Equation 9

$R_s$ = Reflectance white substrate = 0.800

Then, coverage in square feet per gallon is shown by:

Coverage $(Ft^2/gal) = (\text{Factor "A"})(S)$    Equation 10

and consequently,

Coverage $(Ft^2/lb) = \dfrac{\text{(Coverage in } Ft^2/\text{gallon)}}{(8.337)(\text{Sp. Gravity})}$

Equation 11

To determine the film thickness corresponding to the calculated coverage rate,

$$X(\text{Mils}) = \frac{1604.17}{(8.337)(\text{Sp. Gr.})(\text{Coverage in } Ft^2/lb)}$$

Equation 12

This procedure is rather complicated and impractical without the use of a computer.  However, tables have been published which provide all the necessary information (SEE TABLE 2).  Values of Factor "A" for various values of $R_\infty$ and CR are listed.  The user simply locates the $R_\infty$ ($R_w$ in this case) value of the coating over a white background and reads the value of Factor "A" under whichever contrast ratio (CR=0.98, 0.95, or 0.93) desired.  This value is multiplied by "S" and calculations follow through use of Equation 10 through 12.

While this method of determining the hiding power of a coating (i.e., through the use of Factor "A" tables) is the most accurate, if coverage to the nearest square ft/lb is satisfactory, the procedure can be simplified greatly through the use of a Nomograph such as Figure 2.

To determine coverage in square feet per pound of coating material using Figure 2 at CR=0.98 (complete hiding), locate the $R_w$ and "S" values for the coating on the left outermost scales. With a hairline or straight-edge, align these two values. Note the intersection on the center scale (Coverage at Sp.Gr.=1.0). Using this as a pivot point, locate the appropriate specific gravity. Align the two preceding values. The intersection of the outermost right side scale is the expected coverage in square feet per pound. Similar Nomographs can also be constructed for CR=0.95, CR=0.93, etc.

In a similar fashion, the film thickness corresponding to this coverage can be determined from a Nomograph rather than calculations based on Equation 12. Using the Nomograph in Figure 3, the known values for specific gravity and coverage are located on the two outside scales and joined by a straight edge. In the center section, the specific gravity (as used in the left scale) is located and its intersection with the straight edge noted. The intersecting arc corresponding to the film thickness is then followed to the thickness scale on the left. It will be noted that this Nomograph only covers a thickness range of 1-2.5 mils. If the thickness value lies outside this range it can be determined by a slight modification in procedure. For example, if the specific gravity of the coating is 1.38 and the coverage rate is 47.2 square feet per pound, a line connecting these points does not intersect with the center scale of specific gravity at 1.38. This means the thickness is greater than 2.5 mils. To determine the thickness the coverage rate is doubled to 94.4 square feet per pound. A line connecting the specific gravity of 1.38 and 94.4 square feet per pound is drawn and its intersection with the specific gravity of 1.38 is determined. This corresponds to a thickness of just under 1.5 mils--approximately 1.48 mils. Since the coverage rate was doubled, the value for thickness must also be doubled to give 2.96 mils.

Since the values plotted on Figure 3 are independent of "CR" or "S" and equate the simple volumetric relationship between specific gravity, coverage and film thickness, this Nomograph can be used for many other purposes. For example, using the values of specific gravity 1.0 and a film thickness of 1.0 mils, the coverage is determined to be approximately 192 square feet per pound. This, of course, is the theoretical coverage calculated for these values. (Equation 12) Again, using the current example, a material with a specific gravity of 1.38 at 1 mil thickness will cover approximately 138 square feet per pound.

Examples:

Now that the entire procedure has been presented, an example is given to show the relative ease with which the system can be used.

First, the coating weights and $R_g$ and $R_w$ are determined; then, from the Contrast Ratio and Figure 1, the scattering coefficient, "S", is determined. Finally, the coverage rate is determined using either the desired value for Contrast Ratio and Figure 2 or the table for Factor "A" (Table 2) and Equations 10 and 11.

As an example, a typical blue-shade white thin film epoxy powder coating at 40 phr $TiO_2$ is selected. Data are given in Table 1 showing the actual coating weights obtained, the values for $R_g$, $R_w$, and the calculations for "CR" and film thickness. SX is determined using Figure 1 and "S" calculated from Equation 5.

TABLE 1

| | White Glass @8in$^2$ | Black Panel#1 @9in$^2$ | Black Panel#2 @9in$^2$ | Black Panel#3 @9in$^2$ |
|---|---|---|---|---|
| Initial Weight | 40.0980 | 10.4146 | 10.5160 | 10.3018 |
| Weight & Coating | 41.3855 | 10.9399 | 11.0822 | 10.8075 |
| Coating Weight | 1.2875 | 0.5253 | 0.5662 | 0.5057 |
| R $_b$ (Colorimeter) | | 80.9 | 80.9 | 81.0 |
| R $_w$(Colorimeter) | 83.4 | | | |
| CR (R /R ) | | 0.970 | 0.970 | 0.971 |
| X(mils) | | | | |
| (Calculations below) | 7.107 | 2.565 | 2.774 | 2.478 |
| SX (From Figure 1) | | 6.30 | 6.30 | 6.39 |
| S  (SX/X) | | 2.456 | 2.271 | 2.579 |

Average S=2.435
Specific Gravity = 1.383

Calculations for film thickness:

$$X(mils)= \frac{(1000)\ (FILM\ WT\ in\ Grams)}{(AREA)\ (2.54^3)\ (Sp.\ Gravity)} \qquad (Equation\ 3)$$

X (Black 1)    = (4.90) (0.5253) = 2.565 mils
X (Black 2)    = (4.90) (0.5662) = 2.774 mils
X (Black 3)    = (4.90) (0.5057) = 2.478 mils
X (White Glass) = (5.52) (1.2875) = 7.107 mils

NOTE:   Surface Area = 9.00$^2$ for black panels
        Surface Area = 8.00$^2$ for white glass

Using the values from the preceding:

$$R_w = 83.4$$
$$S(Avg) = 2.435$$
$$Sp.\ Gravity = 1.383$$

The coverage characterisitcs as determined using
Figure 2 for CR=0.98 and Equation 12 are:

Coverage Rate in square feet per pound = 47.2
Corresponding film thickness to coverage
rate = 2.95 mils

That is, with a coating system having the
Scattering characteristics as determined
above, a film thickness of 2.95 mils is
required to give complete hiding (CR=0.98).

The coverage characteristics determined using Figure 2
and Figure 3 are:

Coverage Rate in square feet per pound = 47.2
Corresponding film thickness to coverage
rate = 2.96 mils

If the above procedures were not used and the values
were calculated from equations and tables of "FACTOR A", the
following values would have been obtained:

Coverage Rate in square feet per pound = 46.65
Corresponding film thickness to coverage
rate = 2.98 mils

These differences are well within experimental error
and neglible when compared to the reproducibility with which
a coating can be sprayed on a routine basis.  Thus, the
results indicate the experimental coating powder must be
applied at a thickness of 2.9-3.0 mils to give complete
hiding of the substrate.

If a lower degree of hiding is acceptable, the use of
Table 2 for Contrast Ratios of 0.95 and 0.93 shows coverages
of 68.1 and 80.9 square feet per pound respectively,
corresponding to a film thickness of 2.0 mils and 1.7 mils
respectively.

If no consideration is given to scattering, as is
frequently the case in the powder coating trade as it is
today, the coverage for this material based solely on
specific gravity would be quoted as 139 square feet per
pound at 1 mil.

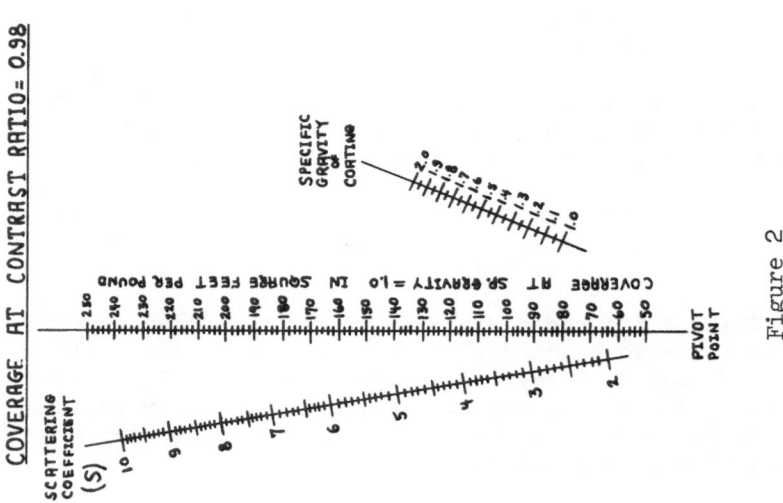

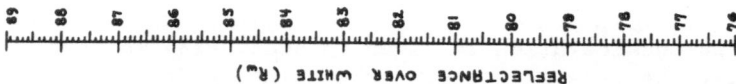

Figure 2

TABLE 2

VALUES OF FACTOR A

| Rw | A.98 | A.95 | A.93 |
|------|--------|--------|--------|
| .810 | 239.00 | 342.03 | 402.65 |
| .811 | 238.23 | 341.19 | 401.80 |
| .812 | 237.47 | 340.35 | 400.95 |
| .813 | 236.70 | 339.52 | 400.09 |
| .814 | 235.93 | 338.68 | 399.25 |
| .815 | 235.17 | 337.85 | 398.40 |
| .816 | 234.41 | 337.02 | 397.56 |
| .817 | 233.64 | 336.19 | 396.72 |
| .818 | 232.88 | 335.37 | 395.88 |
| .819 | 232.12 | 334.54 | 395.05 |
| .820 | 231.37 | 333.72 | 394.22 |
| .821 | 230.61 | 332.90 | 393.39 |
| .822 | 229.85 | 332.09 | 392.56 |
| .823 | 229.10 | 331.27 | 391.74 |
| .824 | 228.34 | 330.46 | 390.91 |
| .825 | 227.59 | 329.65 | 390.10 |
| .826 | 226.84 | 328.84 | 389.28 |
| .827 | 226.09 | 328.04 | 388.47 |
| .828 | 225.35 | 327.23 | 387.66 |
| .829 | 224.60 | 326.43 | 386.85 |
| .830 | 223.85 | 325.63 | 386.04 |
| .831 | 223.11 | 324.84 | 385.24 |
| .832 | 222.37 | 324.04 | 384.44 |
| .833 | 221.63 | 323.25 | 383.65 |
| .834 | 220.89 | 322.46 | 382.85 |
| .835 | 220.15 | 321.67 | 382.06 |
| .836 | 219.41 | 320.89 | 381.27 |
| .837 | 218.67 | 320.11 | 380.49 |
| .838 | 217.94 | 319.33 | 379.71 |
| .839 | 217.20 | 318.55 | 378.93 |
| .840 | 216.47 | 317.77 | 378.15 |
| .841 | 215.74 | 317.00 | 377.38 |
| .842 | 215.01 | 316.23 | 376.61 |
| .843 | 214.29 | 315.46 | 375.84 |

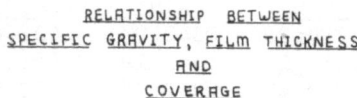

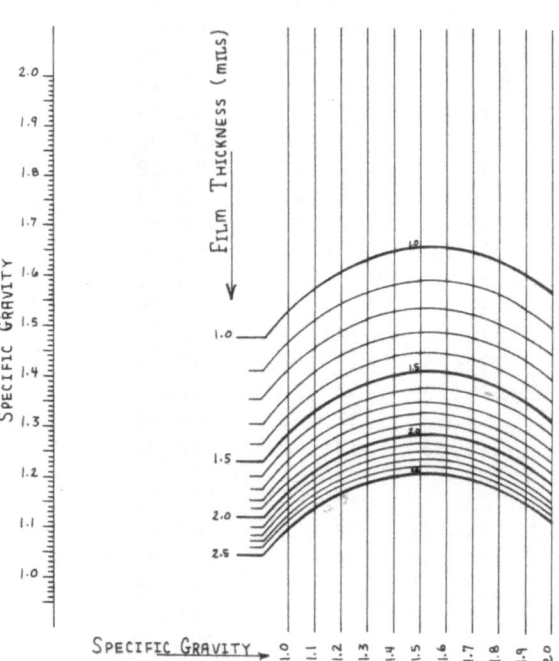

POLYMER CORP.
READING, PA.

Figure 3

## CONCLUSIONS

This procedure, very close to that established in the paint industry, has proven to be simple to use, yet accurate enough to give meaningful results.  Systematic errors are present but to a small degree.  If the same procedure is used for all coatings tested, then the small systematic errors will be cancelled, giving answers which are accurate relative to each other.

References:

1.  P. B. Mitton, J. Paint Tech. 42, 542 (1970)
2.  D. B. Judd, G. Wyszecki, "Color in Business, Science and Industry", J. Wiley & Sons, N.Y. 1967 pp. 387–404
3.  G. Kortum, "Reflectance Spectroscopy", Springer-Verlag New York, Inc., New York, N.Y., 1969, pp. 106–116
4.  D. B. Judd, G. Wyszecki, loc site, pp. 122–130
5.  duPont Pigments, Progress Report No. 4 Revised 1968
6.  duPont Pigments, ibid
7.  P. B. Mitton, A. E. Jacobsen, Off. Dig. 35, 464 (1963)

# FLOW AT THE INTERFACES OF POWDER COATINGS

S. M. Wolpert and J. J. Wojtkowiak

General Motors Research Laboratories

Warren, Michigan   48090

## INTRODUCTION

When a powder coating is baked, its particles simultaneously sinter, coalesce, spread over the substrate and
wick (penetrate) into the microscopic grooves in the substrate.  While these film forming processes occur, the
rough air-coating interface begins to level.  These processes are all interrelated,[1] but a detailed discussion is
beyond the scope of this paper.  Here we are limited to two
phenomena, treated independently:  capillary wicking and
the leveling of powder coatings.

## TIME-DEPENDENT CONTACT ANGLE AND CAPILLARY WICKING

For both spreading on and wicking into a substrate,
the force driving the liquid is proportional to $\gamma \cos\theta$,
where $\gamma$ is the surface tension and $\theta$ is the contact angle,
an equilibrium property.[2]  But how is this driving force
affected if the liquid is so viscous that $\theta$ has not yet
reached its equilibrium configuration?  Is $\theta$ merely
replaced by $\theta(t)$, a time-dependent contact angle?  Or is
the gross shape of the liquid irrelevant to the microscopic
region at the air-liquid-substrate interface?  To investi-
gate this question, a simple model system was chosen:
capillary wicking of a Newtonian liquid into horizontal,
circular capillary tubes.

251

## Theory

Brittin[3] and others[4,5] have thoroughly discussed the capillary wicking equations for liquids of constant $\theta$. Following their arguments but substituting $\theta(t)$, the capillary driving force becomes $f_c = 2\pi R\gamma \cos\theta(t)$. This analysis will be restricted to horizontal tubes with no hydrostatic head. Thus, there is no effect from gravity and $f_c$ is the only force driving the liquid. Wicking is retarded by a viscous force given by the Poiseuille equation, $f_\eta = -8\pi\eta x\dot{x}$, where x is distance of penetration and $\eta$ is the (Newtonian) viscosity coefficient. The contribution from the rate of change of momentum of the liquid, $\dot{p}$, where $p = \pi R^2 \rho xx$, is negligible as shown in Appendix 1. Therefore, $f_\eta = f_c$ and

$$x^2 - x_0^2 = \frac{R\gamma}{2\eta} \int_{t_0}^{t} \cos\theta(t)dt \qquad (1)$$

for the initial point $(x_0, t_0)$. If $\theta$ is constant, then Equation (1) becomes

$$x^2 - x_0^2 = \frac{R\gamma \cos\theta}{2\eta}(t-t_0). \qquad (2)$$

Equation 2 has been verified for many low viscosity liquids whose contact angles reach equilibrium almost instantaneously.[6-9]

## Experimental

Capillary tubes with reservoirs blown or glued at one end were made from "Trubore" precision pyrex tubing as diagrammed in Figure 1 and described further in Appendix B.

The radii were 0.254, and 1.003 mm. They were mounted on a flat glass support on which was glued a strip of millimeter graph paper to serve as a scale for measuring the distance of penetration x. Polybutene-1 oil, $\overline{M_n} = 1500$, Cannon Instrument Company's "Viscosity Standard S-30,000" was introduced into the reservoir by syringe. This fluid is Newtonian at the low rates used here[10] with a viscosity coefficient $\eta(T°C) = 454.9 \exp[-0.30418(T-30°C)]$ poise. Its surface tension was 32.3 dyn/cm by DuNouy tensiometer.

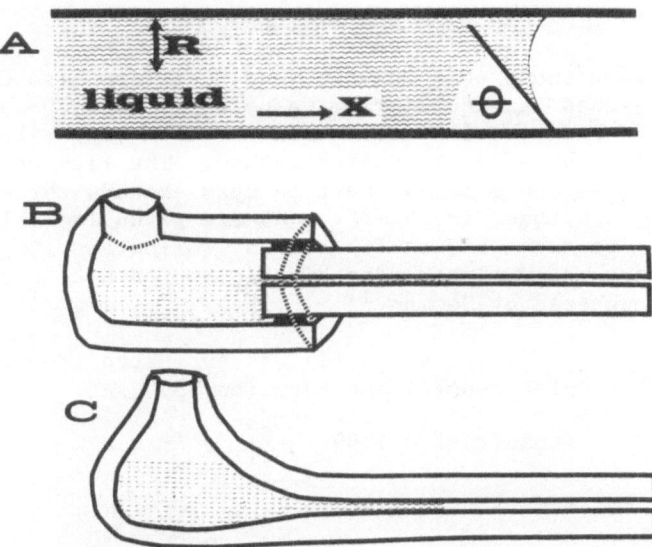

Fig. 1.  Schematic diagram of capillary tubes.  A, reference
diagram.  B, two piece construction with seal.  C, blown
from one piece but position of x = 0 is uncertain to ±0.5 mm.

The traveling meniscus was observed at ambient temperature by low power microscope.  The contact angle on flat
glass was measured with a telemicroscope goniometer but $\theta(t)$
decreased too quickly at lower times for a reliable determination.  Therefore, the meniscus angle $\Phi(t)$ in the tube
was determined from photomicrographs.  $\Phi(t)$ is not quite
the same as $\theta(t)$ because the glass cylinder acts like a lens
distorting the true $\theta$ into the observed $\Phi$.  Comparing them
at intermediate and longer times gave

$$Cos\theta(t) = 1.09 \ Cos\Phi(t) \qquad (3)$$

and this relationship was assumed valid at short times also.
To obtain additional verification of the accuracy of the
conversion factor, the apparent meniscus angle in capillaries
was determined for solvents and solutions whose contact
angles were exactly zero, i.e., for liquids which wet glass
perfectly so $Cos\theta = 1.000$.  Values observed for $\Phi$ ranged
from 4° to 19°.  Although the apparent angles vary substantially from zero, their cosines vary only slightly from
unity, for example, Cos 19° = 0.945.  Therefore, a factor
of 1.09 is reasonable.

### Results and Discussion

Figure 2 shows that $\cos\Phi(t)$, and therefore also $\cos\theta(t)$, has not reached its equilibrium value until over $10^5$ seconds (one day). Polynomial best fits in the form $\phi(t) = \Sigma c_i t^i$ were obtained by a least squares method. The fits were obtained in segments since the time span exceeds three orders of magnitude; the coefficients are given in Table 1. The time-dependence of $\cos\theta(t)$ did not follow any simple function such as a first order kinetic approach to equilibrium as suggested by Newman.[11]

The integral in Equation (1) was evaluated from the polynomial fit for $\cos\phi(t)$ and Equation (3).

$$\int \cos\theta(t)dt = 1.09 \sum \frac{c_i}{i+1} t^{i+1} \qquad (4)$$

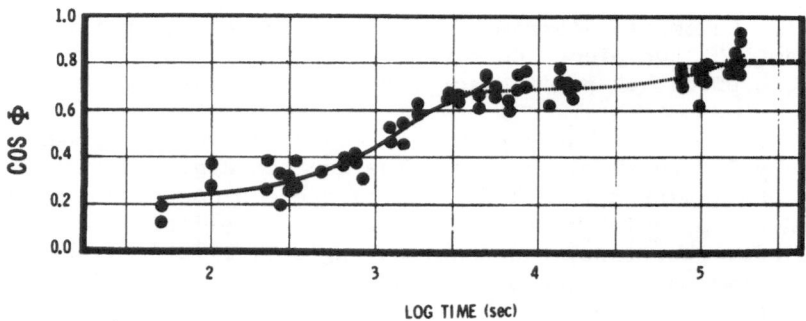

Fig. 2.  Dependence of meniscus angle on time.

### Table 1

Empirical Fit for Meniscus Angle $\phi(t)$

| Time Range (seconds) | $C_0$ $\times 10^{+3}$ | $C_1$ $\times 10^6$ | $C_2$ $\times 10^8$ | $C_3$ $\times 10^{12}$ |
|---|---|---|---|---|
| 0 - 100 | 0.0 | 3320. | | |
| 100 - 4,300 | 213 | 311.4 | -7.84 | 7.42 |
| 3,400 - 150,000 | 679 | 0.873 | | |
| 150,000 - ∞ | 811 | | | |

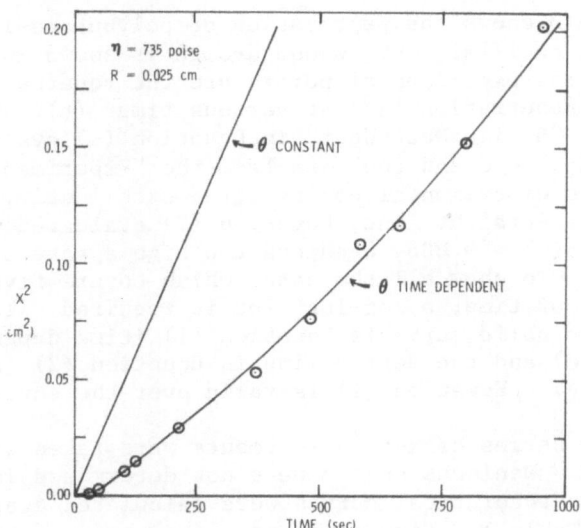

Fig. 3.  Penetration of polybutene-1 into capillary tube.
Linear $x^2$ vs. time.

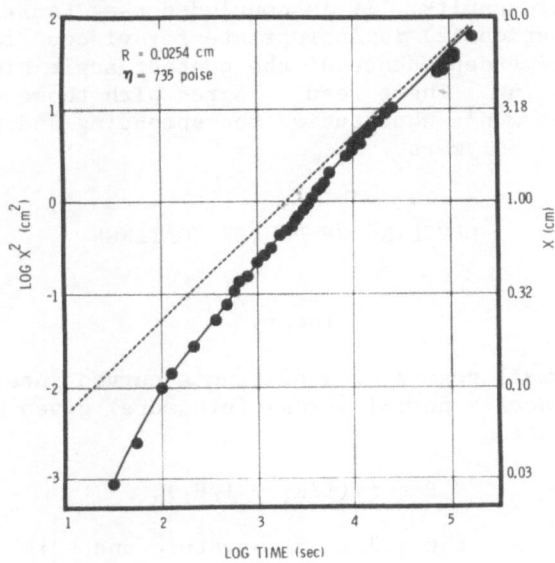

Fig. 4.  Penetration of polybutene-1 into capillary tube.
Same experiment as Fig. 3 but log $x^2$ vs. log time.  Solid
curve, Eq. 1; dotted line, Eq. 2.

Figure 3 shows the penetration of polybutene-1 into a horizontal capillary tube whose design is shown in Figure 1B. The experimental points are the squares of the distance of penetration ($x^2$) at various times (t). The curve labled "θ Time Dependent" is Equation (1) evaluated with x = 0 at t = 0 and the data from the "Experimental" section. The experimental points agree excellently. In contrast, the straight line, Equation (2) evaluated with constant $Cos\theta(\infty)$ = 0.888, predicts too high a rate of penetration. To show all the data, which covers five orders of magnitude of time, a log-log plot is required. In Figure 4, the solid curve is Equation (1) (time dependent contact angle) and the dotted line is Equation (2) (constant contact angle). Equation (1) is valid over the entire range.

Another series of x-t measurements were taken with tubes of design 1C. Meniscus angles were not determined in this series so the theoretical curves were calculated again using the data in Table 1. Figures 5 and 6 show some of the results. The agreement is not as good as was seen before because the zero point is uncertain. Nevertheless, Equation (1) still gives the better fit. Other experiments gave similar results. It is concluded that Equation (1) and not Equation (2) is appropriate for viscous liquids, i.e., the time-dependence of the contact angle significantly affects wicking. These results agree with those of Van Oene, Chang and Newman[12] who studied the spreading and wicking of more viscous polymers.

## LEVELING OF POWDER COATINGS

### Theory

Each small region of a coating's curved interface with air experiences a normal stress (pressure) given by the LaPlace equation,

$$P = \pm\gamma(1/R_1 + 1/R_2),    \qquad (5)$$

where R , R  are the radii of curvature and P is positive for a convex surface. Thus, the surface tension tends to move fluid from peaks to troughs. This mass transport produces a shear stress in the viscous liquid. The hydro-dynamic equations for Newtonian liquids have been discussed

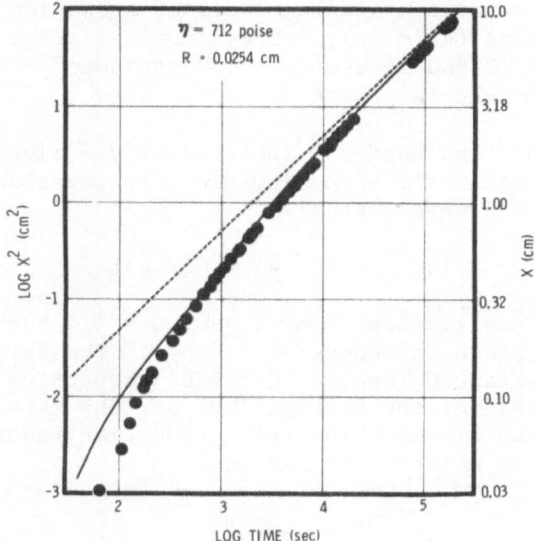

Fig. 5.  Penetration of polybutene-1.  Solid curve, Eq. 1;
dotted line, Eq. 2.

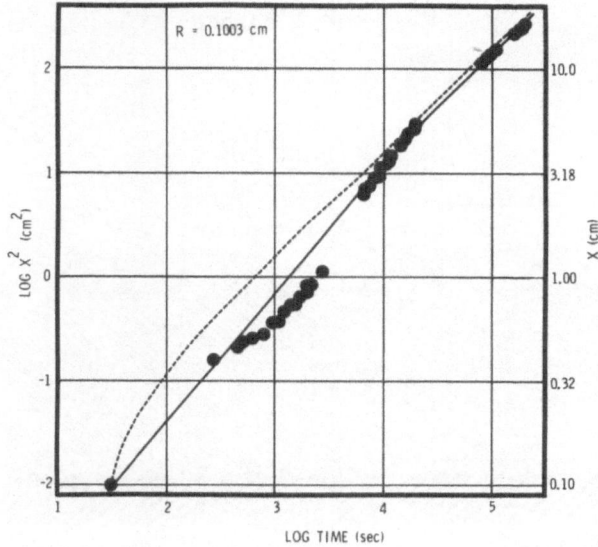

Fig. 6.  Penetration of polybutene-1 into capillary tube.
Solid curve, Eq. 1; dotted curve, Eq. 2.  $\eta$ = 843 poise.

elsewhere for various simple models of a coating's surface: a spherical cap ($R_1 = R_2$);[13] a cylindrical slice ($R_2 = \infty$, $R_1$ constant);[14] and sinusoidal corrugations[15-18] ($R_2 = \infty$) such as diagrammed in Figure 7.

Orchard[15] and Rhodes[17] independently derived the leveling equation for such sinusoidal corrugations of amplitude $\underline{a}$ and wavelength $\lambda$

$$\log \underline{a} = \log \underline{a_0} - Gqf(q)t \qquad (6)$$

where $\underline{a_0}$ is amplitude at time $t = 0$; $G = \gamma/4.606h\eta$; h is the coating's average thickness; $q = 2\pi h/\lambda$. The variables q (dimensionless) and G (sec.$^{-1}$) describe respectively the geometric scale of the corrugations and the material properties of the liquid. The relaxation time required for $\underline{a}$ to decrease by a factor of ten is

$$T = \frac{2.303\eta\lambda}{\pi\gamma\ f(q)} \ . \qquad (7)$$

According to Rhodes[17]

$$f(q) = \frac{(\sinh q)(\cosh q) - q}{q^2 + \cosh^2 q} \ . \qquad (8a)$$

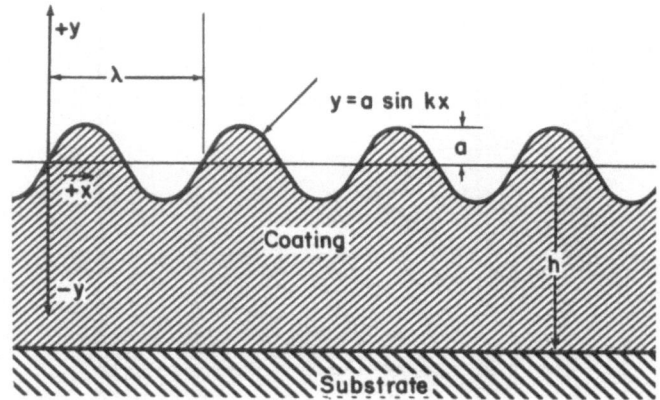

Fig. 7. Reference diagram for model of sinusoidal corrugation, $y = a \sin(kx)$, $k = 2\pi/\lambda$.

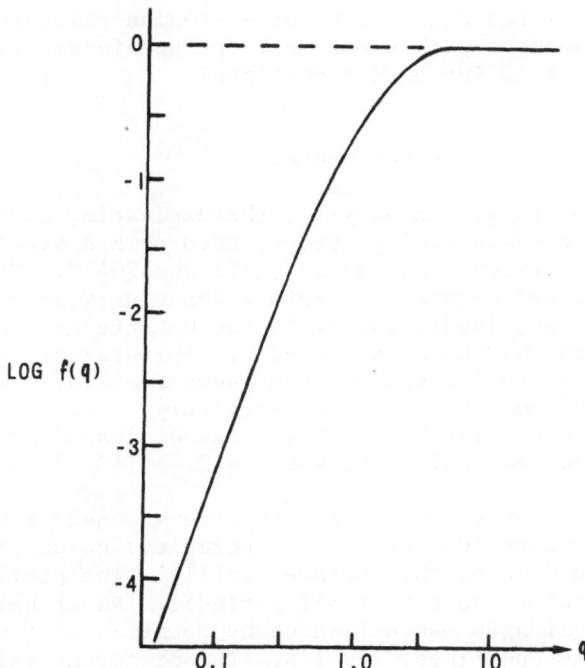

Fig. 8.  Log-Log plot of $f(q)$.

Multiplying numerator and denominator by $sech^2 q$ gives Orchard's[15] expression,

$$f(q) = \frac{\tanh q - q\ sech^2 q}{1 + q^2 sech^2 q} .$$  (8b)

This function is plotted in Figure 8.  (A computer generated table of values is available on request.)  If the film thickness is about equal to, or greater than, the wavelength $(q \geq 2)$, $f(q)$ approaches unity and $T$ does not depend on $h$ at all.  For thinner films $(q < 0.5)$, $f(q)$ approaches $(2/3)q^3$. The relaxation time therefore becomes proportional to $\lambda^4/h^3$, a strong dependence of both wavelength and thickness!

Rhodes verified Equation 6 for Newtonian silicate glasses manufactured with sinusoidal corrugations. Is the equation also valid for powder coatings?


## Experimental

Powder Coatings. An acrylic, thermoplastic, pigmented powder was electrostatically sprayed onto primed steel panels and baked for various times at 177, 191 and 204°C. Viscosities were determined with a Rheotest Rotating Viscometer (Theta Industries, Incorporated) in the Couette mode at shear rates from 0.3 to 0.74 second$^{-1}$. A linear extrapolation to 0 second$^{-1}$ gave, for the above temperatures, 30,000, 12,100 and 5,000 poise respectively. Based on values for polymethylmethacrylate, $\gamma$ was estimated as 30 dyn/cm. The mean thickness was h = 2.56 ($\pm$0.17) mils.

Sections of panel one inch long were scanned with a Gould Surfanalyzer-1200 (Clevite Corporation) to obtain a magnified recording of the surface profile. The profile was very irregular and not at all periodic. Nevertheless, an average wavelength was estimated by counting N, the number of cycles per inch, $\lambda = 1/N$. Two different values of N were obtained depending on whether every peak was counted or only the "major" peaks. Table 2 shows data for the 177°C series. The Surfanalyzer also automatically determined the "roughness", $\bar{A}$, the mean amplitude.


### Table 2

### Thermoplastic Acrylic Powder Coating

| Baking Time t (min.) | Thickness h (mils) | Cycles Per Inch, N | |
|---|---|---|---|
| | | Every Peak | Major Peaks |
| 1 | (5.0) | 50 | 30 |
| 2 | (3.1) | 48 | 22 |
| 3 | 2.7 | 62 | 32 |
| 4 | 2.6 | 50 | 30 |
| 5 | 2.0 | 47 | 27 |
| 10 | 2.7 | 25 | 18 |
| 20 | 2.2 | 25 | 16 |

Sinusoidal Corrugations. The end of a 2" Teflon® rod
was machined into a sinusoidal corrugation of amplitude
a = 6.25 mils (0.00625 in.), and wavelength $\lambda$ = 0.125 in.
A thick (100 mils) film of the same powder as described
above was prepared by casting on flat glass as a suspension
in petroleum ether. The suspension was allowed to evaporate
overnight and coalesced under vacuum at 130°C. The corrugations were impressed into the acrylic film by baking under
vacuum. The Teflon rod was then cleanly snapped off by
immersing it, but not the acrylic film, into liquid nitrogen.

This corrugated film (and its glass support) was heated
for various times at 161°C. Its amplitude and wavelength
was determined after each heating using the Surfanalyzer.
The apparent baking time was corrected for the time required
by the film to reach within 5°C of the oven temperature,
about 130 seconds.

### Results and Discussion

Corrugations. The leveling of the sinusoidal corrugations impressed into the acrylic film is shown in Figure 9.
The data points follow a linear decrease in log a with time
as expected from Equation 6. The dotted line shows the
leveling calculated by Equation 7 with $\gamma$ = 30 dyn/cm,

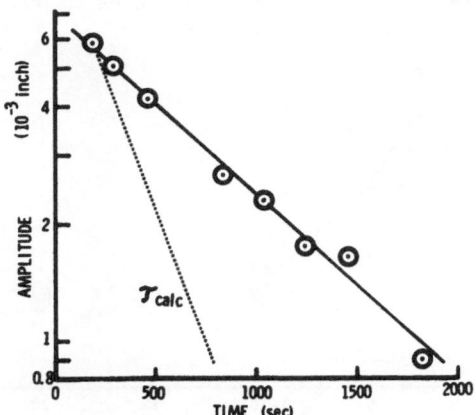

Fig. 9. Decrease in logarithm of amplitude with time for
sinusoidal corrugations impressed into a pigmented, acrylic
film. Dotted line, calculated from Eq. 7.

$\lambda$ = 0.125 inch (observed constant during leveling),
q = $2\pi$(100/125) = 5.0, f(q) = 1.00, $\eta$(161°C) = 92,000 poise.
The observed relaxation time is almost three times that
calculated theoretically.  Additional experiments are in
progress using an unpigmented acrylic polymer.

   Powder Coatings.  The leveling of a typical powder
coating, electrostatically sprayed on to primed panels, is
shown in Figure 10.  The data points show the logarithm of
the "roughness," i.e., the mean amplitude determined by the
Surfanalyzer, after various times of baking at 177°C.  The
line is the theoretical slope calculated from Equations 7
and 8a assuming N = 30, a reasonable value considering the
data in Table 2.  The Orchard-Rhodes equation (Equation 6)
should not be applicable to the data below three minutes
because the powder has not yet coalesced into a continuous
film as evidenced by the anomolously high values for h shown
in Table 2.

   Since coatings behave less like a Newtonian fluid than
a Bingham body, after sufficient leveling the stress from
surface curvature no longer exceeds the coating's yield
stress and leveling stops.[19,20]  This point is reached

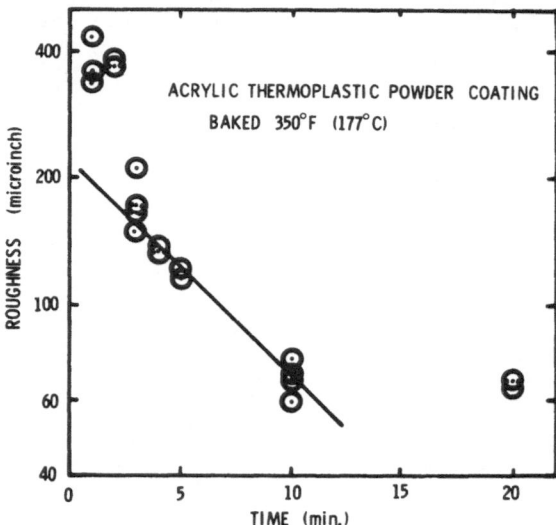

Fig. 10.  Leveling of powder coating baked at 177°C.  Line
calculated from Eq. 7 using $\eta$(177°C) = 30,650 poise,
$\lambda$ = 0.033 in., h = 2.56 mils, $\gamma$ = 30 dyn/cm.

somewhere near ten minutes for this system.  The value of
this yield stress can be estimated from Equation 5, setting
$R_2 = \infty$ and $R_1$ equal to the radius of curvature for a sine
curve $y = a \sin (kx)$ where $k = 2\pi/\lambda$.  Appendix C shows that
under a peak or over a trough,

$$1/R = \pm ak^2. \tag{9}$$

The final amplitude and wavelength were 70 microinches and
0.033 inch respectively.  The yield stress estimated from
this data is somewhere near 33 dyn/cm$^2$.  This rather low
value is nevertheless sufficient to cause an objectionable
"orange-peel" appearance.

Similar results were obtained for the series baked at
the higher temperatures except that since the viscosities
were lower, the relaxation times were shorter as indicated
in Figure 11.  The final roughness decreases with higher
bake temperatures.

An exact comparison of theory with experiment is
impossible with actual coatings because there is no one,
unambiguous wavelength.  To circumvent this problem, the

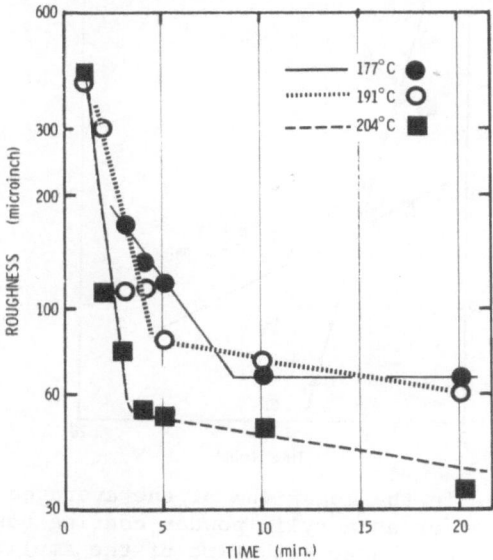

Fig. 11.  Powder coating amplitude, as determined by
Surfanalyzer, as function of time for three bake temperatures

Surfanalyzer output, y(x) was also digitalized in 500 steps of 0.002 inch so it could be subjected to a Fourier analysis.

$$y(x,t) = \sum_{N=1}^{150} C_N(t) \; Cos(2\pi Nx) \qquad (10)$$

The "complex moduli" $C_N(t)$ represent the amplitudes of the profiles' component periodic curves of wavelength $\lambda = 1/N$. For each value of t, a few sets of data were taken and the derived Fourier components averaged to obtain $\overline{C_N(t)}$.

Figure 12 shows representative data for log $C_N(t)$ vs. t for N = 25 and N = 45. The lines were calculated from Equations 7 and 8a. Within the range, 25<N<50, the Fourier coefficients (which represent the coating's real surface) decrease with a relaxation time reasonably close to that derived theoretically. Within this range, f(q) and therefore the expected T depend on N to a power of about 3.2. Such a

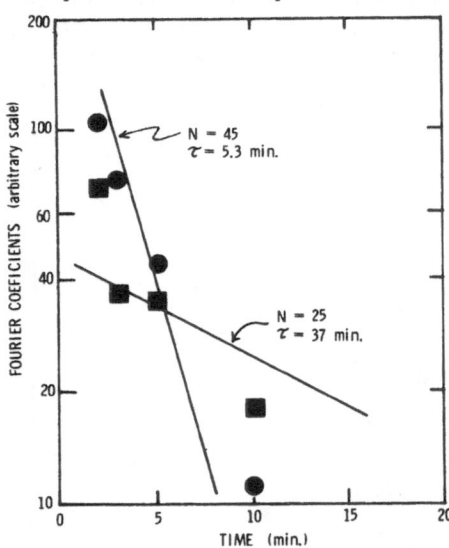

Fig. 12. Change in the logarithm of the averaged complex moduli with time for an acrylic powder coating baked at 177°C. $\overline{C_N(t)}$ represent the change with time of the amplitudes, extracted mathematically, of curves with wavelength 1/N: circles, 0.022 in.; squares, 0.04 in.

rapid change in $C_N(t)$ with N is not suggested by the Fourier results for 25>N>50 (i.e., 0.04 inch<$\lambda$<0.02 inch). However, it is questionable whether the computer program used[21] was giving physically meaningful coefficients for N>50 anyway.

In conclusion, the Orchard-Rhodes equation (Equation 6) describes the data obtained on powder coatings to the correct order of magnitude. More quantitative agreement should not be expected because 1) the actual coating's surface was not a sinusoidal corrugation, and 2) the coating's rheology was complicated by pigments, surfactants, additives and miscellaneous impurities.

## ACKNOWLEDGMENTS

Other members of the Polymers Department deserve our thanks: Mr. D. Hogan for his help in preparing panels and in determining viscosities, Dr. W. Meluch for his advice on computer programming, and Dr. T. Dearlove for his determination of the surface tension of polybutene-1. We also thank Dr. W. Meyer of the Mathematics Department for his advice on Fourier Analysis, and Dr. J. F. Rhodes of the Metallurgy Department for his helpful discussions.

## REFERENCES

1. Van Oene, H., ACS Div. Org. Coatings Plst. Chem., Preprints for Washington D.C. meeting, Sept. 1971, 90-8.

2. Adamson, Arthur W., and Ling, Irene, "The Status of Contact Angle as a Thermodynamic Property," Chapt. 3 in Contact Angle, Wettability and Adhesion, Adv. Chem. Ser. 43; F. M. Fowkes, ed. (ACS, Washington D.C., 1964).

3. Brittin, Wesley E., Jour. Appl. Phys., 17, 37-44 (Jan. 1946).

4. Bosanquet, C. H., Phil. Mag., 45 (6), 525-531 (1923).

5. Pickett, Gerald, Jour. Appl. Phys., 15, 623 (Aug. 1944).

6. Bell, J. M., and Cameron, F. K., Jour. Phys. Chem., 10, 658-674 (1906).

7.  Washburn, Edward W., _Phys. Rev._, _57_ (3), 273-83 (1921).

8.  Peek, R. L. Jr., and McLean, D. A., _Ind. Eng. Chem._, _6_ (2), 85-90 (1934).

9.  Liganza, Joseph R., and Bernstein, Richard B., _Jour. Amer. Chem. Soc._, _73_, 4636-38 (1951).

10. Manning, R. E. (Cannon Instrument Company), private communication (March 1971).

11. Newman, S., _Jour. Coll. Interface Sci._, _26_, 209-13 (1968).

12. Van Oene, H., Chang, Y. F., and Newman, S., _Jour. of Adhesion_, _1_ (1), 54-68 (1969).

13. Rhodes, James F., and King, Burnham W. Jr., _Jour. Amer. Ceram. Soc._, _53_ (3), 134-135 (1970).

14. Patton, Temple C., _Paint Flow and Pigment Dispersion_ Chapt. 5 (Interscience Pub., 1964).

15. Orchard, S. E., _Jour. Appl. Sci. Res._, _A11_, 451-464 (1962).

16. Smith, N. D. P., Orchard, S. E., and Rhind-Tutt, A. J., _Jour. Oil Color Chem. Asscn._, _44_, 618-33 (1961).

17. Rhodes, James F., and King, Burnham W. Jr., _Jour. Amer. Ceram. Soc._, to be published (1972).

18. Rhodes, James F., _Rheology of Vitreous Coatings_, Ph.D. Thesis, Ohio State Univ., 1968.

19. Waring, R. K., _Jour. Rheology_, _2_ (3), 307-14 (1931).

20. Dodge, James S., _Jour. Paint Techn._, _44_ (564), 72-8 (1972).

21. IBM System/360 _Sci. Subroutine Pkg._, III, _Programmers Manual_, "DRHARM" (1970).

22. See for example, "Curvature", in _The International Dictionary of Applied Mathematics_, (Van Nostrand Co., Princeton, N. J., 1967).

## APPENDIX A

From Newton's Second Law, we set the sum of the forces $(f_c + f_\eta)$ equal to the rate of change of momentum with respect to time

$$f_c + f_\eta = \dot{p} = \pi R^2 \rho (x\ddot{x} + \dot{x}^2).$$

(A1)

After rearrangement, one obtains

$$x\ddot{x} + (\dot{x})^2 + bx\dot{x} + c\cos\theta(t) = 0$$

(A2)

where

$$b = 8\eta/R^2\rho$$

(A3)

$$c = 2\gamma/R\rho.$$

(A4)

The solution[3,9] to Equation A2 is

$$x^2 = x_0^2 + \frac{R\gamma}{2\eta}\left[\frac{\exp(-bt)}{b} - \frac{1}{b} + \int_{t_0}^{t}\cos\theta(t)dt\right]$$

(A5)

which is general for any liquid at any time.

Some simplifications are appropriate. If initially the tube is empty, $t_0 = 0$ and $x_0 = 0$. The value of b is estimated by inserting typical experimental data into Equation A3: $\eta(24°C) = 750$ g cm$^{-1}$sec$^{-1}$, R = 0.025 cm, $\rho = 0.85$ g cm$^{-3}$, so $b \approx 10^6$ sec$^{-1}$, a very large number. The two terms which contain b in the denominator are therefore infinitesimal at ordinary times (greater than a microsecond). The simplified Equation A5, still valid for the experimental conditions used here, becomes

$$x^2 - x_0^2 = \frac{R\gamma}{2\eta}\int_{t_0}^{t}\cos\theta(t)dt.$$

(A6)

This result is the same as Equation (1) which was derived merely by equating the capillary and viscous forces.

## APPENDIX B

The two pieces of Design B were glued with clear "Silicone Seal" (General Electric Company). The use of this product did not cause any contamination. To check this possibility, cured Silicone-Seal was extracted overnight with a benzene solution of polybutene of lower molecular weight (Cannon S-8000). Semi-quantitative spectrographic analysis of the filtered solution showed 0.002% Si which is near the limit of detection. Considering the distance from seal to capillary entrance, the low diffusivity in polybutene, and the severity of this test, contamination was not a problem.

Design B had the advantage of a sharply defined starting point but was difficult to seal completely. Design C was blown from a single tube. This simpler design was employed for all runs except the first. It was easy to clean but since there is a neck between the reservoir and region of constant capillary bore extending for about 1 mm, $x_0$ was uncertain. In retrospect, Design B is preferable.

Tubes were cleaned in hot, concentrated chromic acid solution, rinsed several minutes with distilled water, and dried with air which had passed through trains of $CaSO_4$, NaOH-CaO, and activated alumina.

All observations were made at ambient temperature. This temperature was followed continuously with a recording thermometer calibrated to an accuracy of ±0.5°C. The room temperature fluctuated about the mean by ±1.6°C. The mean temperature over the duration of the experiment was used to calculate $\eta(T)$ in Equations (1) and (2).

## APPENDIX C

The radius of curvature, R, for any curve y(x) may be found from[22]

$$1/R = y''[1 + (y')^2]^{-3/2} \qquad (C1)$$

For a sine curve,

$$y = a\mathrm{Sin}(kx), \tag{C2}$$

$$y' = ak\mathrm{Cos}(kx), \tag{C3}$$
$$y'' = ak^2\mathrm{Sin}(kx),$$

where $k = 2\pi/\lambda$.  Substituting,

$$1/R = \frac{ak^2\mathrm{Sin}(kx)}{[1 + a^2k^2\mathrm{Cos}^2(kx)]^{3/2}}. \tag{C4}$$

The Cos term can be neglected if the amplitude is much smaller than the wavelength, which is the case for these experiments.  Directly under a peak or valley, $\mathrm{Sin}(kx) = \pm 1$, so

$$1/R = \pm ak^2. \tag{C5}$$

Recalling Equation (5), the magnitude of the normal stress under a crest is therefore merely

$$\sigma_y = P = \gamma ak^2. \tag{C6}$$

The shear stress, which drives the fluid from peaks into valleys, should be of the same order of magnitude as P.

# INDEX